TRIUMPH STAG

TRIUMPH STAG

The Complete History

James Taylor
&
Dave Jell

Windrow & Greene Automotive Ltd

Published in Great Britain by
Windrow & Greene Automotive Ltd
5 Gerrard Street
London W1V 7LJ

© James Taylor and Dave Jell, 1993

All rights reserved. No part of this publication may be reproduced or transmitted in any form or by any means, electronic or mechanical, including photocopy, recording, or in any information storage and retrieval system, without the prior written permission of the publishers.

A C.I.P. catague record for this book is available from the British Library.

ISBN 1 87200 443 1

Designed by
ghk DESIGN, Bedford Park, London

Advertising Sales:
Boland Advertising & Publishing

Printed in Great Britain by
Butler & Tanner Ltd, Frome and London

THE AUTHORS

James Taylor started researching automotive history in the late 1970s and has developed wide interests in both classic and modern cars. Now devoting himself full-time to writing, he contributes regularly to leading British motoring magazines and to several overseas publications. Among the many books he has written are *Triumph Stag: Choice, Purchase & Performance* and *Triumph Stag Super Profile*. He is also the author of books on Rover (including Land-Rover and Range Rover), Citroen, Mercedes-Benz and Riley.

Dave Jell is the Historian of the Stag Owners' Club. He has owned several Stags, including two of the nine prototypes, and is particularly known for his knowledge of the Stag's pre-production history.

This book is dedicated to the memory of Giovanni Michelotti.

Contents

	Foreword by Harry Webster	6
	Introduction and acknowledgements	6
Chapter one	Ancestry: Triumph and its cars in the 1960s	7
Chapter two	Stag development	17
Chapter three	The Mark I Stag: 1970-1972	39
Chapter four	The Federal Stags: 1971-1973	57
Chapter five	The Mark II Stag: 1973-1975	67
Chapter six	The final Stags: 1975-1977	75
Chapter seven	Post-factory modifications	83
Appendix A	Technical Specifications	93
Appendix B	Vehicle identification	97
Appendix C	Production figures	98
Appendix D	Colour and trim changes	101
Appendix E	Performance figures	107
Appendix F	The Stag Owners Club	111
Appendix G	Bibliography	112
	Triumph Stag Colour Portfolio	113

Foreword

When the Stag was born, it was not part of the Triumph forward 'family planning' programme. However, in keeping with Triumph Engineering intentions of always looking for a niche in the car market not occupied by the mass producers at the time, it filled that requirement with outstanding success.

Without Giovanni Michelotti's styling genius and collaboration it is impossible to say what would have been the fate of Standard-Triumph in the 1950s and 1960s. There would never have been a Stag, and its predecessors from the Heralds through the TRs, Spitfires, GT6s, 1300 and 2000 saloons might not have existed — and certainly not in the same form.

The style of the Stag was entirely Micho's (Giovanni's affectionate name used by all of us at Triumph). He wanted a styling centre-piece for his stand at the Turin Motor Show and chose a Triumph 2000 on which to do it. By prior agreement, it would not go on show if Triumph liked it and wanted to produce it. It was not shown — the rest is history.

It is fitting that this volume should be dedicated to a man of great talent, charm and original design genius — Giovanni Michelotti.

Harry Webster

Introduction and acknowledgements

The idea of a proper history of the Triumph Stag had been developing in both our minds for several years. It had also been developing in the mind of our editor, Bryan Kennedy, and so it was a natural and logical step for the three of us to get together and produce this book.

It was Bryan who approved the outline specification and set the parameters we were to work to; it was Dave who did most of the research, particularly into the early cars; and it was James who did most of the writing.

We hope this book will help to show why the Stag could have been a great car, and why it never lived up to its initial promise. It is quite clear to us that the blame for the Stag's failure should be directed towards those who built it rather than those who designed it: although it did have some design faults when it entered production, these would soon have been developed out in normal circumstances. Sadly, poor manufacturing standards compounded those faults, and circumstances were not normal in the early 1970s. As British Leyland slipped slowly into the red, there was never enough money to revitalise the Stag by putting into production those improvements on which the Triumph engineers were working.

Many people devoted considerable time and effort to providing us with material and photographs for this book. We would like to say here how grateful we are to them. In alphabetical order, then, here they are:

Dennis Barbet, former Triumph development engineer; Steve Barratt, owner of the first production Stag and one of the 4x4 cars; Brian Bayliss, former Triumph design engineer; Gordon Birtwistle, former Triumph development engineer; British Motor Industry Heritage Trust, holders of Triumph archives; Mrs Debbie Bundy, Stag Owners' Club; Mrs. Margaret Colley, wife of the late Harry Colley, Stag Project Engineer; Martin and Peter Cox, former Triumph development engineers; Nigel Cross, former Triumph development engineer; Dave Destler, editor of *British Car* magazine; Tony Hart, proprietor of Hart Racing Services, Stag specialists; Heathrow Stag Centre, Stag specialists; Mark Johnson, Northants Stag Centre; Spen King, former Triumph Director of Engineering; Tony Lee, former Chief Experimental Engineer at Triumph; John Lloyd, former Triumph Chief Engineer and later Director of Engineering; Edgardo Michelotti, son of stylist Giovanni Michelotti; Peter and Marilyn Robinson, Registrars of the Stag Owners' Club; Harry Webster, former Triumph Director of Engineering and prime mover behind the Stag project.

We should also acknowledge that *Classic Cars* magazine published an early version of Chapter Two in its February 1992 issue.

James Taylor
Dave Jell

CHAPTER ONE

Ancestry: Triumph and its cars in the 1960s

During the 1960s, Triumph almost went bankrupt, lost its independence in a takeover and then, as the result of later mergers, found itself belonging to the same manufacturing group as many of its former rivals. During the next decade, the Triumph name was dragged through the mud along with the rest of the British Leyland marques and was displaced from the large saloon car market by Rover. In 1981, Triumph sports cars went out of production and by 1982 Triumphs had ceased to exist altogether except as rebadged, licence-built Hondas.

Into the middle of this turmoil the Triumph Stag was born, a car which did not fit into the established Triumph market areas and which attempted to break new ground. No car, however good, could have succeeded in such circumstances; and the Stag, while a first-rate design in most respects, suffered from prolonged teething troubles.

When work began on the Stag in the mid-1960s, Triumph had long since ceased to be an independent motor manufacturer. The original Triumph Motor Company, founded in Coventry to manufacture bicycles in the 1880s, had gone into receivership in the late 1930s. The War held up resolution of its problems, but shortly before peace came, Triumph was bought out by Standard. Although Standard mechanical components soon came to figure in cars with the Triumph badge, the company's owners perpetuated and built on the marque's sporting reputation, and during the 1950s the TR sports cars proved a massive success, especially in export markets. So high was the Triumph flag flying, in fact, that the marque name was applied to small Standard saloons as a way of selling them in the United States and, when the time came for them to be replaced, their successors were badged from the beginning as Triumphs.

This huge success with the TR2, TR3 and TR3A sports cars in the 1950s was one of the reasons why Standard-Triumph planned for a major expansion during the following decade. After disentangling itself from an agreement with the Canadian Massey-Harris company, who had bought the rights to the Ferguson tractor which Standard-Triumph had been making under licence since 1946, the company began to invest in new plant: first with a new assembly hall on the main Canley site just outside Coventry, and then with a vast new factory at Speke, near Liverpool. These would permit the introduction of new models and an increase in production of existing ones, but Standard-Triumph was also anxious to be self-sufficient. Starting in 1958, when it purchased the coachbuilders Mulliners of Birmingham in order to give itself a body-manufacturing capability, it continued to buy up component suppliers throughout the decade.

However, just as Standard-Triumph's reserves were fully stretched by this expansion programme, so the British economy took a downturn. In order to prevent inflation in 1960, the British Government raised interest rates, which meant of course that hire purchase and credit became more expensive. This took the edge off car sales at home, and in the meantime sales in the United States had failed to live up to expectations. By spring 1960, life was not looking good for Standard-Triumph. Cash flow deteriorated; by November, the company was seriously in the red and the situation was getting worse by the day. Standard-Triumph was losing £600,000 a month and, with prospects continuing to look bleak, thousands of production-line workers were put on to short time. Over the winter of 1960-61, a three-day working week was introduced.

Fortunately, however, the bus and truck combine Leyland Motors was looking to expand its interests into the car-manufacturing field and, seeing the ailing Standard-Triumph company, quickly stepped in with a takeover bid. The bid was made in November 1960; the agreed merger was made public at the beginning of December and, although Leyland almost pulled out when it discovered that Standard Triumph had lost £3 million since September that year, the deal went ahead and was formally concluded in April 1961.

Most of the Standard-Triumph directors were removed the following August, and the new Leyland-owned company mounted a big export drive for the 1962 season with its new TR4 sports model. This drive was a roaring success: production went up from a nadir of 78,383 in the 1961 season (of which fewer than 3,000 were TRs) to 100,764 for 1964 (of which more than 9,000 were TRs), and the figures were to go on climbing for the rest of the decade. Only Government domestic policies caused a slight and temporary downturn for the 1967 season but, by then, further changes in the ownership of the company were on the horizon.

The 1960s were the decade of the merger for the British motor industry, and the Leyland takeover of Standard-Triumph in 1961 was simply the first in a long line. BMC (itself formed by the amalgamation of Austin and the Nuffield Group in 1952) made the next important move, by buying up Pressed Steel, who built the bodies for several major British car manufacturers. This forced Jaguar into the security of an alliance with BMC (which then became British Motor Holdings, or BMH) and Rover into the similar security of an alliance with Leyland.

By 1967, therefore, there were two major groupings in the British motor industry alongside American-owned or -controlled Ford, Vauxhall and Rootes Group. On one side was British Motor Holdings, and on the other was Leyland. Fearing further erosion of national ownership of the motor industry, and arguing that these two major groupings would effectively be cutting each other's throats in export markets, Prime Minister Harold Wilson encouraged a shotgun marriage between BMH and Leyland. The result, early in 1968, was the creation of British Leyland.

The merger left Standard-Triumph in the uncomfortable position of facing in-house competition in all three of its product areas. On the big saloon front, it was now up against Rover; in the sports car market, it was threatened by MG; and its upmarket small saloons were now being crowded by Riley, Wolseley and the Vanden Plas derivatives of Austin/Morris models. No doubt many Triumph people therefore pinned their hopes of the marque's very survival on its planned new Stag, a grand tourer which faced no competition from any other British Leyland product.

Triumph models of the 1960s

In the 1960s, anyone who had been asked what the Triumph name stood for would undoubtedly have replied 'sports cars', although in fact the Triumph range was far more diverse than that. It was true that the company offered 'traditional' sports cars in the TR range and smaller sports cars in the shape of the Spitfire and its GT6 derivative; and it was true that it put up some spirited performances in the prestigious European motor races, such as the Le

Ancestry: Triumph and its cars in the 1960s

Above: *In the 1960s, Triumph's reputation as a sports car manufacturer depended heavily on cars like this four-cylinder TR4.*

Below: *By the end of the decade, the TR4 had been transformed into a TR6, with a fuel-injected six-cylinder engine.*

Opposite page
Top: *Triumph was also successful at the cheaper end of the sports car market, with the four-cylinder Spitfire...*
Bottom *... and adding a six-cylinder engine and a fastback made the Spitfire into a GT6.*

Right: *Right at the bottom of the saloon car range was the Herald, of which over 500,000 were built during a twelve-year production life. (BMIHT/Rover Group)*

Mans 24-Hour Race. But its larger saloons were simply well-finished executive-class models with above-average performance, and by no stretch of the imagination could the small Herald, with its Courier van derivative, be described as sporting. Triumph seasoned the mix with more sporting variants, of course — the 2.5 PI in the case of the larger saloons and the Vitesse in the case of the Herald; but the undeniable truth was that not every 1960s Triumph was a sports car or even a 'sporting' car.

None of the model ranges of the 1960s could trace its ancestry back further than 1952. In that year, Triumph announced a new sports model with which it intended to cash in on the roadster market which MG had so successfully opened up in the USA. That car, which was of very simple design and used mechanical components derived from the Standard saloon ranges, was the TR sports model. Productionised as the TR2 (the original TR1 design exhibited at the Motor Show had been considerably modified), the car's all-enveloping styling stole a march on MG, who persisted with old-fashioned wings and running-boards styling until 1955, and initiated Triumph's sales successes in the vast American market. By the mid-1950s, sales of Triumph TRs in the USA greatly exceeded sales at home, and this brought about a major switch in the company's orientation. From now on, what the US market wanted was more important to the company than what the home market wanted. Thus, when America wanted more power and better brakes, it got them in the 1956 TR3 and, by the end of 1957, export deliveries of TRs had increased by over 100 percent. And the figure was to continue rising until 1960 saw a downturn.

Nevertheless, Triumph bounced back in 1961 with the restyled TR4, improved its suspension in 1965 with the TR4A, added a six-cylinder engine in 1967 for the TR5 (called TR250 in the USA), and restyled the range yet again for 1968's TR6. Throughout the decade, it was TR sales in the USA which were primarily responsible for keeping Triumph's accounts firmly in the black.

As far as the small saloons were concerned, the models of the 1960s had originated in 1959. Small saloons from the Standard-Triumph combine during the 1950s had all been Standards since 1953, badged as 8s or 10s and later also bearing the Pennant name. However, the positive image which Triumph had earned in the marketplace with the success of the TRs had persuaded Standard-Triumph to rebadge these little saloons as Triumphs for the US market, where the Standard 10 became a Triumph TR10 in 1957, and Standard-Triumph decided that the replacement small saloon should wear Triumph badges from the start.

Thus, the Triumph Herald replaced the small Standards. In spite of a rakish, twin-carburettor, coupé alternative, its 948cc made it nobody's

Left: *The separate-chassis, rear-wheel-drive Herald was later supplemented by the monocoque, front-wheel-drive 1300 saloon. (BMIHT/Rover Group)*

Below: *The six-cylinder version of the Herald was the Vitesse, in production from 1962 to 1971. (BMIHT/Rover Group)*

Opposite page
Top: *Triumph's large saloon was the 2000, and it was from an early example of this that the original Stag prototype was made.*
Bottom: *In 1969, the 2000 range was facelifted, using styling elements associated with the still-secret Stag. This example is a later 2000TC.*

sports car and its swing-axle rear suspension gave it handling most unbecoming to a sporting model. However, an increase in engine size to 1147cc in 1960 did wonders for sales, the introduction of a more highly tuned 12/50 version two years later gave the model a further boost, and the arrival in the intervening year of a 'small six' 1.6-litre derivative called the Vitesse firmly established the small Triumphs in the market. Both Herald and Vitesse would be further up-engined in the second half of the decade (the Herald to 1296cc in 1967, the Vitesse to two litres in 1966).

The success of the Herald/Vitesse family prompted Triumph to introduce a second small saloon range in 1965, this time using monocoque construction and front-wheel drive instead of the Herald's separate chassis and rear-wheel drive. The 1300 was another success story, and Triumph ex-panded on its appeal in 1967 with the introduction of the quicker 1300TC companion model.

Meanwhile in 1961, the Herald had spawned a fourth range of cars. It was in that year that Triumph took the newly up-engined 1147cc Herald chassis, modified it slightly and fitted it with an attractive two-seater sports body. The Spitfire, always intended as a rival to the MG Midget, fed off the TRs' sporting image and became a major success in its own right. Successive models gained more performance, the tricky swing-axle rear suspension was mercifully tamed in 1968, and in 1966 the Spitfire also spawned a companion model with the two-litre engine, called the GT6.

The final Triumph range of the 1960s — which made the total up to five — was the executive saloon range. In the Standard-Triumph scheme of things, these had actually replaced the Phase III

Standard Vanguard saloons, but with the change to Triumph badging had come a change of orientation, and the 2000 models were aimed at the new two-litre 'executive car' class. Introduced in 1963, they had monocoque shells and the six-cylinder engine which would later see service in the GT6 and two-litre Vitesse. Strong sales were supplemented by an estate car derivative after 1964, the sporting overtones were emphasised with the fuel-injected 2.5 PI version after 1968, and a particularly successful facelift in 1969 ensured that sales of these cars would remain strong until well into the 1970s.

During the 1960s, then, Triumph's model range was already complicated for a relatively small company. There were large and small sports car ranges, and large and small saloon ranges, with the small saloons being sub-divided into two types during the second half of the decade. Nowhere among them was a grand tourer, and nowhere in Triumph's post-1945 history had there ever been one. Yet, at one of the least auspicious times in its entire history, Triumph chose to develop one in the shape of the Stag.

The people behind Triumph

In spite of all the changes during the 1960s, the key people at Triumph mostly remained in place. At the top of the tree were the Chairman, Sir Donald Stokes (who was also Leyland's Managing Director and Deputy Chairman), Chief Executive George Turnbull, and Director of Engineering Harry Webster, who was also effectively his own product planner for most of the decade. All had

Above: Stag bodies were built at the 'Liverpool no. 2' factory at Speke on Merseyside, and were then delivered to Canley for final assembly.

Opposite page
Over 10,000 people were employed at Triumph's Canley headquarters, and it was here that Stag development took place. Design and development were carried out away from the administrative headquarters and assembly halls, in the cluster of buildings at the top right of this picture.

Right: Harry Webster, Director of Engineering until he moved to Austin in 1968, was the prime mover behind the Triumph Stag.

been on the Triumph Board since the Leyland takeover in 1961 (although Stokes had not become Chairman until 1963), and there would be no changes until the Leyland-BMC merger of 1968, when Harry Webster moved to Austin at Longbridge and was replaced at Triumph by Spen King, from Rover.

Triumph, however, had no Chief Stylist, even though it did have a styling department which was ably run by Les Moore. That had been the position since 1955 when Walter Belgrove, who had been in charge of Triumph styling since the 1930s, resigned from the company after a row with Deputy Managing Director Ted Grinham. Standard-Triumph had struggled on for a while, but had not engaged a new Chief Stylist by the time it came into contact with the Italian stylist Giovanni Michelotti in 1957.

Although Michelotti's name is now inextricably linked with Triumph products, his own story went back much further. After gaining experience with Ghia-Aigle in Switzerland, he set up as an independent designer in Turin in the 1940s. The early 1950s saw him working closely with the coachbuilder Vignale on some appealing bodywork for Ferrari, and it was mainly this which established the reputation he enjoyed by the middle of the decade.

Michelotti's meeting with Standard-Triumph was brought about by the businessman Raymond Flower, who approached the company about supplies of mechanical components for a new car he wanted to build and sell in Egypt. Quite by chance, he happened to mention that he knew where he could get cars styled and built in a matter of only two or three months. As Standard-Triumph was having considerable trouble with new car designs at the time, notably with the successor to the Standard 8 and 10, it eagerly followed up this lead. The swift stylist proved to be Michelotti, the swift builders Vignale of Turin.

Alick Dick (Triumph's Managing Director in pre-Leyland days) and Martin Tustin (then General Manager) asked Michelotti to draw up a sports car for them, and were highly impressed with the 'Dream Car' he constructed on a TR3 chassis in 1957. Before long, they had him on a retainer as Standard-Triumph's consultant stylist. Michelotti reworked the stodgy Vanguard Phase III shape to make the 1959 Vignale Vanguard, drew up some further TR proposals, and from then on styled every new Triumph introduced until the end of the 1960s — the Herald, the Spitfire, the 2000, the TR4, the TR5 and the Mark II 2000. Had he not been too busy, he would also certainly have completed the restyle for the TR6 at the end of the decade. It was Michelotti too, as we shall see in the following chapter, who drew up the design which would enter production in 1970 as the Triumph Stag.

The 1960s were an age in which personalities dominated Triumph. So it was that when the consultant stylist came up with a design which the Director of Engineering liked, it did not take long for the whole company to convince themselves that it would be a Good Thing. And when the product planners got to work on it, they had little alternative but to confirm the decision which had already been made to put it into production!

Triumph factories

Although the administrative headquarters of Triumph was at Canley, near Coventry, the company in fact owned several factories which were spread over a wide area. The Canley headquarters had belonged to Standard since 1916, and since 1946 its factory buildings had also housed Triumph production lines. In 1961, a large new assembly hall had been completed on the site, to cope with the new model ranges planned for the decade.

Canley, however, was primarily an assembly plant. During the 1960s, the bodies for Triumph cars were actually built at the company's Bordesley Green factory, near Birmingham (the old Mulliners/Forward Radiator Company plant), or at the Liverpool factory acquired during 1960. And from 1969, bodies also came from the new Speke (or 'Liverpool no. 2') factory.

Body preparation was undertaken at the old Fisher and Ludlow plant in Tile Hill, near Coventry, before delivery to Canley. Suspension and steering components arrived from Alford and Alder, in Hemel Hempstead, while heavy castings came from the old Bean Industries factory in the Staffordshire town of Tipton, and other components from the Radford plant in Coventry.

CHAPTER TWO

Stag development

In the early 1960s, Michelotti enjoyed an excellent relationship with Triumph, both professionally and on a personal level with their Director of Engineering, Harry Webster. So when he asked Webster if he could have an old Triumph 2000 to turn into a show special, his request was willingly granted. But there was one proviso: that Triumph would have first claim on the show design if the company liked it.

A rather tired 2000, registered 6105 KV, did its last duties for Triumph as a service support car at the 1965 Le Mans 24-Hour Race and was afterwards driven straight to Michelotti's studios in Turin. Webster then probably thought no more about it until, on one of his regular visits to Turin in the summer of 1965, he caught sight of the part-finished show car. It was completely unrecognisable as having Triumph 2000 ancestry, not least because it sat on a wheelbase shortened by six inches and had been conceived as a two-plus-two convertible. There were quad headlamps hidden behind power-operated sliding grille sections, and the sculpted front and rear shapes were remarkably attractive. Webster liked it at once and offered to buy the completed car for his company. As a result, the Michelotti design never did appear on a show stand, but instead became the styling prototype of a new Triumph.

While the 2000 special was being completed at Michelotti's studios in late 1965 and early 1966, Harry Webster talked to other Triumph men about the car. He formally reported to his fellow-Directors during February, observing that he thought it looked more promising than the TR4A facelift which was then under consideration. Indeed, for a few weeks, the car was referred to as a TR6, although it rapidly became clear that the Michelotti design was too much of a grand tourer ever to slot comfortably into the TR sports car range.

Over the next few months, Triumph soberly examined the car's sales potential, but it was already clear that the Board was sold on the idea of producing it. A letter Harry Webster wrote to Michelotti on 21st April 1966 discussed design modifications in terms of 'when' rather than 'if', and a formal Product Proposal was put before the Triumph Board on 7th July. As Webster himself put it at the time, his department had never been so enthusiastic about a new model as they were about

THE MICHELOTTI METHOD

Initially an artistic sketch of a proposed new style was presented. After discussion and modification, and probably many sketches later, a one-tenth size layout was made, and when this was considered satisfactory a full-size layout commenced.

On a vertical drawing-board marked out in 10cm squares and big enough to take a full-size side elevation and front and rear end views, the initial outline was made with chalk. Surprisingly, white chalk lines showed up quite well on the white paper and were easily altered as necessary by erasing with a duster and redrawing as required until the desired form was agreed.

At this point, the free-hand chalk lines were lined-in in heavy black pencil, using the usual draughtsman's sweeps and curves, the form at each 10cm interval longitudinally and laterally then being established.

From this information, a carpenter's shop made plywood templates of each 10cm segment, again lengthwise and laterally, and these were then assembled in their correct positions to give a full-size car body in 'egg-box' form. Any necessary 'mods' were made at this stage to arrive at the desired shape.

The finished buck was then clad in metal by skilled panel-beaters and tinsmiths to make a complete body structure which was subsequently painted and trimmed. Mechanicals were then fitted.

From first line to running prototype took some three to four months, and the operation was relatively cheap even at that time. Such has been the subsequent level of inflation that it is difficult to imagine that Michelotti charged just £10,000 *total* for the first three Herald prototype bodies — a coupé, a saloon, and an estate car.

Harry Webster

Opposite page
The original Triumph 2000-based Show car takes shape in Michelotti's Turin studios, late in 1965 or early in 1966.

Above: *The finished car outside Michelotti's premises, looking much longer and lower than the production cars would, thanks mainly to its smaller (13-inch) wheels. Note that the fuel filler was on the left-hand-side.*

this one, a state of affairs which he saw as a good augury.

By the time the Product Proposal went before the Triumph Board, the new project had acquired the name of Stag (Triumph favoured four-letter code-names for its projects in the 1960s). It was already clear that the car would take the Triumph marque into a new market sector, and that this would entail certain risks. Nevertheless, Triumph's sales people believed that the market for luxury sports cars was then increasing at the expense of that for small ones, with the result that the Stag would take the company in the direction the market was heading. The alternative way of meeting this changing demand, as Harry Webster spelled out to the Board at its July meeting, was to give the TR4 a major face-lift, for which the tooling and other expenditure would amount to nearly as much as it would cost to put the Stag into production.

But it was Sir Donald Stokes, Triumph's Chairman and Managing Director of Leyland Motors, who weighed-in most heavily in favour of the Stag. He told the Board at that July meeting that, if the company was to survive, it should set trends rather than follow them. For this reason, he said, conventional market surveys based on past experience were not entirely satisfactory. He believed the Stag would strengthen the sports car franchise which was the basis of Triumph's business and would therefore put the company in a much stronger competitive position. The Board agreed with him and authorised Harry Webster to commit resources to preparing the Stag for production. The speed with which this commitment had been achieved is remarkable, especially since there was then only one prototype in existence, and that was little more than a styling exercise!

It *was* a viable road car, however, even though it would need development. One problem was insufficient weight at the rear end (perhaps because

Opposite page
Top: The Michelotti car had sliding headlamp grilles, which normally covered the headlamps like this.
Bottom: Electric motors moved the outer grille sections behind the centre section when the headlamps were needed.

Right: Michelotti's original interior design was not much liked at Triumph.

the original 2000 saloon springs had been retained with the lighter body), and the boot had to be loaded with a hundredweight of ballast before Harry Colley — a senior Triumph engineer who later became Stag Project Engineer — thought it was safe enough to drive back to Canley from Turin in the summer of 1966. In spite of this rather makeshift modification, Colley was impressed with its handling. The exact date of his arrival with the car in the UK is unknown, but by June the Michelotti prototype was being examined in Triumph's Styling Studios.

Triumph's most important market in the late 1960s was the USA, so it was not surprising that the first concern was how to sell the car there. Anticipated new safety regulations prompted Harry Webster to ask Michelotti in a letter dated 11th July to style a roll-over bar for the car, which he called the 'GT 2000', presumably to clarify things for Michelotti who would not have been aware of its new Stag code-name. He also wanted to have a hard top; he wanted to discuss the styling of the hood and interior; and he wanted to discuss headroom in the rear. Other documents show that Triumph had built their own interior styling buck and had found rear headroom could be improved by lowering the seat pan, while lowered footwells would give more legroom.

It was already clear that Triumph were looking on this car as a product which would expand their traditional sports car market. With it, they hoped to produce a European-style grand tourer which would have all the image of a Porsche or an Alfa Romeo, but would cost considerably less; and in particular, they wanted to capture a slice of the luxury sports car market which was then domi-

Left: Triumph planned to turn the Stag into a Mercedes SL competitor, and they bought an example of the German car to examine closely. In October 1966, it was pressed into service as a tow-car to take the Michelotti prototype back to Turin for modifications to be made. Here it is en route through Switzerland; the numberplate of 6105 KV is just visible under the tarpaulin.

Opposite page
Top: Michelotti added a roll-over bar, which did not initially meet with Triumph approval...
Bottom:... and a hard-top which was, frankly, ugly.

nated by the Pagoda-roof Mercedes-Benz SL. To that end, they bought an example of the current 230SL to study, and in due course deliberately copied several elements of its design. One of these was the combination of hard and soft tops which the German car was then the only one to offer.

The Mercedes was eventually stripped down and studied at Canley, but in the meantime it was pressed into service to tow a trailer bearing the Stag prototype back to Turin for Michelotti to carry out modifications to suit Triumph's wishes. This was in October 1966, and it was probably in early November that a delegation of senior Triumph people went out to Italy to see what progress had been made. Along with Harry Webster went his deputy, Chief Engineer John Lloyd, Triumph's Chief Executive George Turnbull, and Sir Donald Stokes. However, they expressed a 'strong feeling of disappointment' with what they saw, singling out for criticism the seats, hood, roll-over bar and hard top. As Harry Webster put it in a letter to Turin, 'the result of your work over the last few weeks is not really Michelotti'; and he pointed out that the Italian need not hesitate to tell Triumph if the company's requirements conflicted with his own aesthetic judgement.

Triumph requested 22 further modifications to the car in November 1966, and Michelotti agreed to carry them out as quickly as possible. 'We are,' as Webster observed towards the end of November, 'completely held up here, pending return of the vehicle.' In fact, the car did not return to Canley until the beginning of January, and then without a modification to the door pillars which Triumph had wanted and which Michelotti had warned them would take a long time to effect. Chief Body Engineer Arthur Ballard wrote to Michelotti on 5th January, noting that the car was now 'in most respects satisfactory', but that Triumph wanted to raise the rear of the hard top, to improve both the appearance and the headroom. Michelotti was asked to submit appropriate drawings.

Canley takes over

Once these modifications had been made, however, Michelotti's work was done. Now it was up to the Triumph designers and engineers to get to work on the Stag concept, and in fact the Italian seems to have been involved with the project very little over the next two years.

As built, the Stag prototype had the 1998cc OHV six-cylinder engine of the 2000 saloon,

Above: Triumph also asked for interior changes, so Michelotti stripped out the seats and dashboard…

Left. … and tried again. However, Triumph were still not pleased with the result and eventually designed their own interior for the car.

THE STAG PROTOTYPES

6105 KV. Michelotti Styling Prototype
Converted from Triumph 2000 saloon in 1965 by Michelotti. Finished in lilac/pink. Later used for development of hard top and roll-over bar. Converted to fastback in autumn 1968. Never allocated an X-series (Triumph experimental) number, probably because the 2000 from which it was converted already had its own chassis number.

X 763. First Triumph Prototype
Probably completed in October 1967. Finished in matt black with red interior. Originally fitted with 2½-litre, six-cylinder carburettor engine; later had 2½-litre carburettor V8, then fuel-injected 2½-litre V8. Eventually fitted with three-litre V8. Used for pavé, braking and rear axle tests. May never have been road-registered. Scrapped in mid-1970s.

X 777, PVC 237G. Second Triumph Prototype
Build completed on 28th March 1968. LHD with automatic transmission. Originally had 2½-litre six-cylinder engine, later 2½-litre V8 (initially with carburettors and later with fuel injection); finally fitted with three-litre V8. Originally pale yellow. Fitted with air conditioning and US marker lamps, also with fuel expansion tank for US safety-regulation experiments. Dual exhausts and hidden headlamps. Roll-over bar from new; first car fitted with T-bar, during 1968. Fitted with CV joints instead of UJs in rear suspension. Seen in early pictures with number plates GDU 680D (front) and 1325 HP (rear). Used for pavé testing, air conditioning tests in North Africa, brake tests in North Wales (October 1969), and suspension tests. Survives.

X 782, PHP 465G. Third Triumph Prototype
Built circa November 1968 as RHD automatic, but converted to manual by July 1970. Initially fitted with 2½-litre carburettor V8, then 2½-litre fuel injected V8, and finally three-litre V8. White. Used for engine and gearbox work, including emissions control development. Scrapped by Triumph on 8th September 1978.

X 783, TKV 754J. Fourth Triumph Prototype
Built circa November 1968 as RHD automatic with air conditioning. Initially fitted with 2½-litre carburettor V8, then 2½-litre fuel injected V8, and finally three-litre V8. Originally Valencia Blue. Used for 50,000 mile engine test and for US-market styling development. Survives, now registered as 3702 DF, but the owner intends to restore its original registration number.

X 787
Bodyshell only. Sent to Michelotti in Turin, March 1969, for conversion to fastback. Built up in UK as X 798.

X790, RHP 659H. Fifth Triumph Prototype
US-specification development car, built with three-litre V8. Wedgwood Blue. To USA in spring 1970; used for cooling and air conditioning tests at Pikes Peak. Fitted with Bosch fuel injection by October 1970.

X 798. Sixth Triumph Prototype
Built from bodyshell X 787, modified by Michelotti to fastback. Dark blue. Completed by February 1970, with fuel-injected 2½-litre six-cylinder engine. By September 1971, running with 2½-litre carburettor six-cylinder engine and 13-inch wheels. Almost certainly never registered. Scrapped on instructions from Triumph at a Nottingham scrapyard.

X802. Seventh Triumph Prototype
LHD car built with three-litre V8. Barrier crash test at MIRA, 15th December 1969. Subjected to front barrier crash test, rear barrier crash test and side impact tests on both sides. Scrapped after roof intrusion test on 18th March 1971. Never registered.

X 815, WHP 852J. Eighth Triumph Prototype
Built as fastback by Triumph, circa February 1971. Fitted with three-litre V8 from new. Originally Saffron Yellow, later White. Survives, re-registered as PAE 755, after a careful and sympathetic rebuild.

Above: Michelotti's second attempt at a roll-over bar and hard top was much more successful. Note the Standard-Triumph emblem on the upright of the new roll-over bar. The steel wheels must have been fitted in Turin, but Triumph put wires (with two-eared spinners) back on the car when they examined it at Canley.

which put out 90bhp at 5000rpm. This was adequate, but it was not going to give the sort of performance which Harry Webster wanted in production cars. Much more appropriate would be the long-stroke version of the engine then under development for the TR5 sports car, which promised between 130 and 150bhp when fitted with Lucas fuel injection. This would give very respectable performance for the first production cars and then, at some future date, it could be supplemented or replaced by the new V8 engine which was under development.

The V8 was actually part of a new family of engines on which Triumph's Chief Technical Engineer, Lewis Dawtrey, had been working since 1963. The aim was to save both tooling and development costs by developing an overhead-camshaft slant-four and a V8 which would in effect be two slant-fours on a common crankshaft. One or the other would be suitable for all Triumph's cars, for the capacities available in theory ranged from the 1½ litres of the smallest four-cylinder to four

Opposite page
Top: The first Triumph V8 engines were very different from the eventual production type. This is the block of one of the later 2½-litre engines, with the water pump in its eventual production position. Comparison with a 3-litre block readily makes apparent the smaller bore size.
Below: The second Triumph-built prototype was completed in March 1968 but not registered (as PVC 237 G) until several months later. In this picture, it is wearing false number plates.

litres in the largest feasible V8. There was no doubt that a powerful, modern V8 engine would give the Stag considerable appeal in its main target market of the USA, and Harry Webster decided to try one of these engines in a Stag prototype as early as he could.

Rover, meanwhile, which had only just joined Triumph in the Leyland empire, had already tooled-up to put their ex-General Motors 3½-litre V8 engine into production. As Leyland management swiftly recognised, it therefore made no sense for Triumph to continue to spend money on the development of their own V8, which would eventually prove to be a direct competitor for the Rover engine. As a result, Harry Webster was instructed to try out a Rover V8 in the Stag. The trial was conducted during May and June 1967, and must have been done either on a mock-up or on the Michelotti prototype, as this was then the only Stag in existence.

At this stage, of course, the Triumph and Rover engineers still saw one another as rivals — they had been united under Leyland for only a matter of months — and the Triumph engineers seem not to have tried too hard to make the Rover engine fit the Stag's engine bay. At the beginning of June, Harry Webster told the Triumph Board that the height of the Rover engine was causing installation prob-

lems; and not long after that, the project appears to have been pronounced impossible. It was not, of course: Triumph already knew that the Rover engineers were looking at fuel-injection systems which would reduce the overall height of their V8 enough to make it a snug fit in the Stag's engine bay. Similarly, it needed only a modified inlet manifold (such as was developed for the MGB GT V8 in the early 1970s) to reduce the height of the carburetted engine sufficiently. One way or another, the Triumph engineers' unwillingness to persuade the Rover V8 to fit under the Stag's bonnet left the way clear for further development of their own V8. Leyland, either through ignorance or inability, did not attempt to interfere further.

While this off-stage manoeuvring was going on, other engineers at Triumph were concentrating on getting the styling right, as the body press tools would have to be signed-off as the first stage in getting the car into production. It was November 1967 before final agreement was reached; but from then on, things began to happen more quickly. As their usual suppliers were already heavily committed on other projects, Triumph commissioned the body tooling from Karmann of Osnabruck, a well-respected German coachbuilding company to whom they had recently turned for the TR5-to-TR6 restyle when Michelotti had also proved too busy.

In the last months of 1967, Triumph built up their own prototype Stag, almost certainly using a modified 2000 saloon floorpan as its basis in exactly the same way as Michelotti had done with 6105 KV. By the end of the year, X 763 was on the road, running a 2½-litre six-cylinder engine but with carburettors, no doubt because the Lucas fuel injection had not yet been developed to a satisfactory state. A second prototype, X 777, was completed in March 1968, this time with left-hand-drive. This also started life with a six-cylinder engine, but the dual exhausts visible in photographs taken in July 1968 suggest that it was by then fitted with a V8.

The first prototype Triumph V8s seem to have been ready by the late spring or early summer of 1968, and replaced the fuel-injected six-cylinder engines in the two existing Stag prototypes. So keen were the engineers to see what the V8-engined Stag would be like that, rather than wait for the correct components to become available, they cobbled up a makeshift exhaust system for the first car and ran it over the roof! Their enthusiasm was rather misplaced, however, as these first V8s proved to be far from satisfactory.

Known by its project code of PE 158, the original Triumph V8 had a capacity of 2½ litres. It was tried out initially with carburettors and later with fuel injection, first a modified Lucas system and then a Bosch type. The original carburetted engines had their water pumps driven from the

Above: Stag Project Engineer Harry Colley went with the second Triumph-built prototype to North Africa in order to conduct some air conditioning tests. The black marks on the body sides of the unbadged car are temperature sensors: one of the tests involved leaving the car to stand in the hot sun for several hours and then switching on the air conditioning to see if it could cope!

Opposite page
This is the second Triumph-built car again, with twin exhausts suggesting it had a 2½-litre V8 engine by the time of this July 1968 picture. Note that the soft top has no quarter-lights and that the fuel filler is now on the right-hand rear wing.

Below: During the autumn of 1969, the hard-worked PVC 237 G went on an endurance run in North Wales. Here it is undergoing brake testing in typical scenery.

jackshaft at the front of the engine, but later examples had the water-pump built into the vee between the cylinder banks, partly so that the metering unit for the fuel injection system could be run from the jackshaft. Early examples also had a mechanical fuel pump located near the bottom of the engine and operated by a long lever; and, in spite of the original plan to commonise parts between the new four-cylinder engine and the V8 wherever possible, the cylinder heads of the 2½-litre V8 were not interchangeable with those of the four-cylinder engine.

Testing with the first Triumph-built prototype, X 763, soon revealed some serious structural shortcomings. There was 'the most enormous scuttle-shake,' as Harry Webster told a motoring journalist some years ago. 'You almost had to try and catch the steering wheel, if you know what I mean!' Neither double-skinning in the body sills nor a TR-style saddle support between facia and floor could cure the problem, and it was left to John Lloyd to demonstrate (with the aid of a broom-handle!) that the only really effective solution was to brace the roll-over bar to the windscreen header rail. The second Triumph prototype, X 777, was modified to incorporate the new T-brace during the summer of 1968, and X 763 was also modified very soon afterwards.

Related development work seems to have been carried out on other cars. As Triumph wanted to put their V8 engine into the 2000 saloon in due course, a number of V8-powered 2000 development cars were built. In order to perfect the sliding headlamp covers incorporated in Michelotti's original design, at least one 2000 (a red car) was also fitted with these. However, things were not destined to run smoothly on the Stag project during 1968.

It was in that year that the Leyland Motor Corporation (which owned Standard-Triumph) merged with British Motor Holdings to create the new British Leyland Motor Corporation, and the spring months were disrupted by the consequent reorganisation. Harry Webster was moved from Triumph to take over the volume car division of

Above: When Triumph decided to investigate a fastback body style for the Stag, the long-suffering Michelotti prototype 6105 KV went back to Turin for further work. Here it is being rebuilt in Michelotti's workshop.

Opposite page
The third Triumph-built prototype was pictured on 4th November 1969 undergoing testing at Mallory Park. The driver was Gordon Birtwistle.

Austin-Morris, and in his place as Triumph Chief Engineer came Spen King, one of Rover's most gifted engineers.

King rapidly got to grips with the Stag project, where the 2½-litre V8 had run into difficulties. The engine lacked low-speed torque when fitted with carburettors and could not be made to meet American emissions control regulations when fitted with fuel injection. King therefore suspended the work with fuel injection and asked the Triumph engineers to redevelop the engine with a bigger bore. At the Board meeting on 5th July 1968, the first one King attended as Triumph's Chief Engineer, he confirmed that the PE 188 three-litre V8 would replace the PE 158 2½-litre type, and that a choice between carburetted and Bosch petrol-injected forms for initial production would be made on the basis of cost, availability, and efficiency results from the forthcoming test programme.

In view of the problems which that three-litre V8 caused the Stag during its later production life, it is interesting to examine its design here. From the outset, it had two weaknesses. The first was the position of its water pump, mounted so high up in the vee between the cylinder heads that even a slight drop in water level would leave the pump quite literally high and dry, pumping no coolant around the engine. In the controlled conditions of the Triumph experimental department, it worked perfectly well, but in the real world where owners took less care over the maintenance of their vehicles, it proved disastrous. The second weakness was its tendency to blow head gaskets, which had already been apparent when the 2½-litre V8s were built. The origin of this problem dated back even further. The primary cause was the angled cylinder head bolt design adopted at Saab's request for the related four-cylinder engine when Triumph had developed a version of this for the Swedish company to use in its new 99 model in the mid-1960s. These engines did give some trouble in service, but in the interests of minimising development costs, the Triumph engineers were instructed to use the same angled-bolt design for the V8. Saab in fact reverted to the original Triumph design with the bolts at right-angles to the block when they took over manufacture of the engine themselves in 1972; but the V8 and all the later Triumph slant-fours related to it retained the Saab head bolt design with its associated problems.

Financial and time considerations prevented Triumph from changing the head bolt design when the three-litre V8 was under development in 1968. As it was, the engineers had to find a cheaper solution to the engine's head gasket problems. They settled on a German-made rubberised head gasket, which gave a positive seal between head and block and was in every way superior to conventional types. However, it was also much more expensive, and for that reason was abandoned. The only other way around the problem was to use conventional gaskets and torque the heads down more tightly, but this was never satisfactory. There

Stag development

Opposite page
The finished product was not very satisfactory, however, and was rejected. In this view, it is clear that Michelotti had also experimented with an alternative 'egg-box' design for the grille.

Below: *To slim the side elevation, Michelotti had blacked out the sills, and he had also blacked out the tail panel. Triumph would do the same on production Stags between 1973 and 1975. When the car returned to Canley, Triumph replaced the rear bumper with a standard production type and added five-spoke wheel trims, which improved the car's looks considerably.*

were constant head gasket failures during development, and these earned the Stag two telling nicknames: 'the Mobile Kettle' and 'Colley's Folly', the latter after Project Engineer Harry Colley.

The head gasket problems were further compounded in later years by manufacturing quality problems. As Spen King remembers, the prototype V8 engines had been cast by a first-class foundry outside the British Leyland group, but he was pressurised into giving the production contract to a Leyland-owned foundry. 'Their quality standards were just awful,' he says, 'but nobody would listen to me. BL said manufacture had to be kept within the company, and that was that.'

King's misgivings were well-founded: in production, an alarmingly high percentage of cylinder blocks had to be rejected, and quality control was so bad that some completed engines were found still to contain casting sand. This undoubtedly contributed to a number of premature engine failures.

Meanwhile, the change from 2½-litre to three-litre V8 during development had set the Stag programme back by some three months. It also had a considerable knock-on effect. Triumph had

originally been intending to use a number of components from existing production cars in order to keep development costs and development time to a minimum. But the extra torque from the enlarged V8 demanded both a stronger gearbox and a stronger rear axle, and Triumph also decided to go for bigger brakes with larger wheels to accommodate them. All these took time to develop and added to the costs.

The sliding headlamp arrangement was a further casualty over the summer of 1968, as it could not be made reliable, and the third Triumph-built prototype probably had exposed lamps from the beginning. The restyled front end which resulted was in fact so attractive that Triumph asked Michelotti to graft it onto the forthcoming

Opposite page
Michelotti built up the chosen design in metal on the bodyshell, which was then shipped back to Triumph and built up as a complete car.

Below: *Triumph asked Michelotti to try again with the fastback design, and this time sent him a spare bodyshell. Here it is with two alternative designs mocked-up in clay for assessment.*

Innsbruck (2000 and 2.5 PI Mk.II) saloons, together with a matching, Stag-like rear end; and these saloons entered production a whole year before the Stag which had influenced their styling.

With the third and fourth Stag prototypes, which were built together and completed some time around November 1968, Triumph's new grand tourer was gradually taking on its definitive form. The six-cylinder engine was no longer under consideration, and both of these cars were built with 2½-litre V8s. Within a few months, their engines and the 2½-litre V8s in the two earlier Triumph-built prototypes were replaced by new three-litre V8s. At this stage, Triumph management was hoping to have the car ready for announcement at the Earls Court Motor Show in October 1969, but various component supply difficulties and further development problems had put paid to that hope by the spring of 1969. And in the meantime, the designers had started to look even further ahead, and to look at an additional variant of the Stag for possible future introduction.

The fastbacks

Exactly why Triumph decided to look at the possibility of a fastback Stag is not clear, but an earlier plan to build a 2000 fastback had been abandoned in favour of the 2000 estate, and it may be that fond memories of that project still lingered on. Whatever the reason, the fastback Stag project seems to have had a fair head of steam behind it by October 1968, when Michelotti was asked to convert 6105 KV into a fastback coupé. The car went out to Turin for a third time, Michelotti performed one of the rapid design-and-build jobs for which he was famous, and the car returned to Canley for viewing during November.

As usual, Triumph wanted to tinker with the design, and at the end of January, Spen King sent Michelotti some pictures of a modified design which had been drawn up at Canley. Michelotti was unimpressed and, in mid-February, he submitted drawings of an alternative design. The upshot was that on 20th March, 1969, a hand-built prototype bodyshell — logged as number X 787 in Triumph's prototype records — was sent out to Italy and Michelotti was asked to turn it into a new fastback. Whether the style he adopted after trying two different ones in clay on the shell was primarily his own or Triumph's is unclear. It certainly was different from the style used on 6105 KV, however, giving the Stag the appearance of an oversized GT6 from

Left: The last Stag prototype was a third hatchback, styled by Les Moore and his team at Triumph and built up at Canley early in 1971.

some angles. The rebuilt shell was back at Canley by 22nd August, when it was propped up on blocks in the styling studio and mocked-up with wheels for initial viewing.

Renumbered as X 798, the fastback shell was subsequently built up as a complete car. Curiously, it was fitted with a fuel-injected six-cylinder engine, even though this had long since been rejected as a production option. One explanation might be that all the V8s then in existence had been earmarked for the launch stock of production cars and that none could be spared for development vehicles. Nevertheless, the fastback retained its six-cylinder engine as late as September 1971, when twin carburettors had replaced the fuel injection and the car was running on 13-inch wheels instead of the 14-inch ones which had come in with the three-litre V8.

Production begins

As far as the European-specification cars were concerned, development was complete by the middle of 1969; construction of the first pre-production car (numbered LD 1) began in July and was completed in November. As far as it is possible to tell, no model-name had yet been proposed for the production cars, but it was probably around this time that the Stag was shown to Triumph's US dealers as a forthcoming new model. The US dealers liked the project name of Stag so much that they persuaded Triumph to retain it for the production cars.

Development now concentrated on the US-specification models, which Triumph planned to introduce a year or so after the European launch in the first half of 1970. Two further prototypes were constructed for final development of the US specification. One — X 802 — was simply a crash test car which was subjected at MIRA to the barrier, side and roof intrusion tests then demanded under Federal law. The other — X 790 — was a full US-specification car which went out to the USA in spring 1970 and was used for cooling and air conditioning tests at Pikes Peak. By October 1970, its three-litre V8 engine had been fitted experimentally with Bosch D-Jetronic fuel injection, no doubt because this newly-introduced system looked like offering the most effective way of meeting emissions control requirements. The fact that Lucas planned to manufacture it under licence in the UK may also have had something to do with Triumph's interest, but in fact no fuel-injected Stag ever did go into production.

For the future, Triumph seem to have been thinking of more power and of the fastback as an alternative model. The completed fastback (X 798) was among the cars at a viewing arranged for the

Triumph Directors on 11th February 1970, and a four-valves-per-cylinder V8 was under consideration as early as December 1969. This would have had a version of the four-valve cylinder head which Spen King and Lewis Dawtrey had designed for the Dolomite Sprint's slant-four engine, which was of course effectively half a V8. Detailed estimates prepared by November 1970 showed that the four-valve V8 would weigh some 20 lbs more than the production two-valve type, and would give 225bhp with peak torque of 212.5 lbs/ft. Sadly, the engine never even reached the prototype stage. 'It was simply pushed aside by the pressure of other work,' remembers Spen King.

One other project for coaxing more power out of the existing engine also came to a head in the summer of 1970. The plan in this case was to replace the existing crankshaft with a single-plane type, and an engine with the experimental crankshaft was installed in X 763, the first Triumph-built prototype. Gordon Birtwistle road-tested the car, and remembers that the engine was free-revving but had vibration problems. Its exhaust note had also lost the characteristic rumble of a Stag V8 and, adds development engineer Dennis Barbet, there were no power gains. Work with the single-plane crankshaft went no further.

Many people at Canley must have been holding their breath during the first six months of 1970. The Stag launch had been scheduled for June, in Belgium, but by March there were still no production cars. Strikes and last-minute development problems had caused hold-ups, with the result that just two pre-production cars had been built: LD 1, completed in November 1969, and LD 2, a left-hand-drive car mocked-up to US Federal specifications when it was completed on 12th February 1970.

Production proper finally got under way on 13th March, and the first few Stags came down the TR6 assembly lines (they never would have a dedicated assembly line, for the usual cost reasons). At first, they came down the line singly or in twos and threes; but as assembly built up, they began to come down in larger batches. Meanwhile, pre-production car LD 1 was photographed during March for the launch publicity and the first sales brochure. As Triumph was keen to give the Stag a European image which would help to establish it as a competitor to the Mercedes-Benz SL and Alfa Romeo 1750 models, these pictures were taken in Italy, Switzerland and the south of France.

The launch went ahead without a hitch; the press reviews were generally enthusiastic; and for the first few months, Triumph's main concern was with matching production to demand. On the development side, there was still more than enough to do to get the car ready for its US-market launch in September 1971. And there was still the question of the fastback variant to consider…

Fastback again

Although the second Michelotti-built fastback was a vast improvement on the first, Triumph still had some reservations. Once the pressure to get the basic Stag into production and to get it ready for the US market had eased a little, the company turned again to the fastback. A new design was drawn up by Triumph's own stylists under Les Moore, and a third hatchback prototype, numbered X 815, was built up in the early part of 1971 and put on the road in February or March of that year.

'I liked it,' Spen King recalled some eighteen years later. 'I thought it was a nice, sensible car, and much more useful than an ordinary Stag.' With further hindsight, however, it is easy to see why the fastback was never approved for production. Firstly, it arrived just as the Stag was beginning to cause Triumph a major headache with warranty claims relating to the engine. Secondly, it coincided with the growing realisation that the British Leyland honeymoon was well and truly over. All the corporation's profits were being pumped into the volume cars division, leaving nothing for the development of prestige models like the fastback Stag. X 815 was therefore the last Stag prototype to be built.

Of the nine Stag prototypes, only three are known to survive. The second Triumph-built car, X 777, is currently owned by SOC Spares, the Stag parts specialists, but is in very poor condition. The fourth Triumph-built car, X 783, survives in private ownership; and the final hatchback prototype, X 815, has been restored and is owned by a Stag enthusiast. There are rumours that one other hatchback survives, but that is all.

The Stag had not been announced, of course, when LD 1 went down to the south of France for publicity pictures in March 1970. In this picture, Mr Holmes of the Triumph publicity crew with the car is removing the identifying badges from the rear wings in an attempt to maintain secrecy. The Triumph name on the bumper-mounted plinth has already been taped over.

CHAPTER THREE

The Mark I Stag: 1970-1972

It is worth reviewing the specification of the Stag as settled for production in the summer of 1970. The basic configuration of the car as a V8-engined 2+2 had, of course, been settled much earlier, but the 'showroom specification' was agreed much nearer the actual launch date. Considerations of pricing undoubtedly influenced the decision to offer three transmission options: four-speed manual or three-speed automatic, with overdrive an extra-cost option on the manual cars. Similarly, there would be three basic body configurations, the cheapest having only the soft top, the next most expensive having only the hard top, and the most expensive version of the car coming with both hard and soft tops.

As Appendix D shows, there were also six paint options (all taken from the existing Triumph range) and four interior colour options which, together, made up a total of 13 possible variants. The total number of Stag variants theoretically available, including body, mechanical, paint and trim options was therefore 117 (13 colour/trim combinations x three body styles x three transmissions) — which rather gives the lie to the commonly-held view that 'all Stags are the same'! At this stage, however, there were only a very few optional extras, and these were limited to items like radios and wing mirrors.

There were nevertheless some undeniably strange elements in the Stag's specification, and there can certainly be little doubt that cost had once again been the major influence on them. Thus, the seats were upholstered in vinyl, when leather would have been much more fitting for a car of the Stag's pretensions (it was in theory an extra-cost option, but no cars are known to have had it from new). The soft top had to be erected by hand, when a power-operated top would have been more appropriate (and had, in fact, been tried out during development). The hard top was inordinately heavy and cumbersome ('I wanted a lightweight one in aluminium,' says Spen King, 'but I was over-ruled'). And, lastly, the wheel covers were the same imitations of five-spoke alloy wheels which Triumph had applied to so many of its models in the late 1960s. As if their appearance was not enough to suggest they were cheap accessories, they actually had five imitation wheel-nuts when the wheel underneath was held on by only four real ones!

Initial reactions

The Stag was introduced to the British motoring press at Knokke-le-Zoute, in northern Belgium, early in June 1970. Representatives from some of the leading European and American magazines also seem to have been present, although the car would be available strictly on the home market for the immediate future. Triumph's aim in inviting these other journalists was presumably to gain publicity before the actual launches outside the UK.

In fact, the Stag was not the only car which the press were seeing for the first time in Belgium. The press launch also introduced the front-wheel-drive 1500 and rear-wheel-drive Toledo medium-sized saloons. However, the Stag was undoubtedly the star of the show. A dozen cars went out to Belgium, numbers LD 4 to LD 14 and LD 17, bearing registration numbers RVC 425 H to RVC 435 H and RVC 438 H. Ten had overdrive, two were automatics, and three wore hard tops. It also looks as if pre-production car LD 1 (RRW 97 H) might have been in Knokke for the press launch. Nevertheless, the press fleet proper, established in July, consisted of just six cars: RVC 429 H to 431 H, RVC 434 H and 435 H, and RVC 438 H.

All the launch cars had been tested at MIRA to ensure they were capable of 120mph — 2mph more than the sales catalogues claimed as the Stag's maximum speed — and all were re-tested on the Jabbeke straight (a section of the Brussels to Ostend motorway) so that the two fastest cars could be identified. These were then given to *Autocar* and *Motor*, the two most influential publications represented at the launch, in the hope of generating favourable press comment.

Autocar, *Autosport*, *Car* and *Motor* reported their first impressions in the next available issue. All reported favourably, although they did criticise certain details. Charles Bulmer, in *Motor* of 13th June, wrote that 'without any doubt this is going to be a very successful car and perhaps the source from which even more successful variants may spring.' For *Autosport*, John Bolster summed up the car's refined character with the words, 'the Stag is as far as it could possibly be from a hairy sports car'. And *Autocar* took a global view, seeing it within the context of British Leyland rather than purely Triumph products, and observing that 'from the British Leyland point of view, it plugs the gap between the MGB and the Jaguar E-type.'

Opposite page
The very first production Stag was used in all the early publicity photographs. This one dates from the time of the car's public announcement in June 1970 and shows that it had several differences from the production models which followed. The sills have a trim strip (not introduced in production until 18 months later), the radio aerial is on the offside front wing (it was more commonly fitted to the rear nearside), and there is no BL logo on the front wings.

Below: *This later picture shows the car with the BL logo in place, as stipulated by British Leyland for all 1971-model Triumphs. (MRP)*

Generally, the press liked the ride and the layout of the dashboard and controls, found the performance more than adequate, and appreciated the efforts which had gone into noise suppression. However, wind noise over 60mph, with either the soft top or the hard top in place, came in for criticism, and the power-assisted steering was another source of dissatisfaction, being rather too light. Nevertheless, the critics conceded that longer acquaintance helped to overcome initial concerns about the steering.

Autocar was subsequently lucky enough to get its hands on a car for long enough to carry out a full road test, and published the results in its issue dated 30th July. Broadly speaking, its initial impressions were confirmed: this was a refined and satisfying long-distance tourer which was enjoyable to drive fast. But the magazine's testers were unhappy with its wet-weather behaviour, claiming that it could be tail-happy and that they had even managed to unstick the front wheels on a wet corner. This criticism did not come as a total surprise to at least one Triumph engineer, who had been responsible for the skid pan tests of the car. He remembers having doubts about the wet-road performance of

Left: *The launch cars in the press garage at Knokke, in Belgium. RVC 435 H was LD 14 O, a left-hand-drive overdrive model finished in Saffron with a Tan interior. It was later used in the James Bond film* Diamonds are Forever, *although the overdubbed engine noise on the soundtrack was that of a Triumph Herald! The identity of the lady in the picture is not clear.*

the Michelin XAS asymmetrical-tread tyres with which all Mark I British Stags were fitted at the factory.

Autocar also reported that the Stag supplied for test (RVC 431 H) had been unable to match Triumph's claimed maximum speed of 118mph. The magazine recorded a mean maximum of just 115mph on the MIRA test track, and pointed out that the car accelerated no faster than a 2.5 PI saloon. Triumph test driver Gordon Birtwistle was given the task of taking the car back to MIRA to verify this failing, and he concluded that *Autocar*'s testers had not tried hard enough! Triumph therefore gave the magazine another opportunity to check the car's performance, but nothing appeared in print until nearly a year later, when *Autocar* still managed only 116mph on test.

An exception to the favourable reviews which the Stag received came from *Motor Sport*, whose Bill Boddy was rather miffed at not getting his hands on a test example earlier on and made some rather more serious criticisms in the magazine's October issue. He considered the steering's absence of 'feel' to be quite unacceptable for fast motoring, disliked the manual gearbox, and thought the Stag's acceleration left something to be desired. But he did concede that the car was attractive to look at and reasonably priced.

On sale

Although the Stag was theoretically available in Triumph showrooms in Great Britain immediately after the press launch, many dealers refused to sell the examples they had until production built up and they were assured of getting further stocks. This was to prove a wise move, as Stag production actually stopped completely for some three weeks in August and September 1970 while Triumph ironed out some quality control problems.

Nevertheless, some cars were sold to the public before the Earls Court Motor Show in October, and probably around 30 were registered with the H-suffix registrations current until the end of July. Stags were in short supply at first, however, and this can hardly have pleased Triumph's customers. Moreover, the price had gone up by £160 by the time the car went on display at Earls Court. One angry customer wrote to *Motor* magazine, which published his letter in its issue dated 17th October. He had ordered his car on 9th June, the day it was announced, and was still waiting for it to be delivered. The main distributor with whom he had placed his order had by then received cars only for onward transmission to dealers; 'so how,' the letter-writer asked, 'can a company justify such a price increase when the car has only been available for

Right: In the press garage at Knokke again. Mark I Stags all had a courtesy light on each pillar of the roll-over bar, as can be seen here.

16 weeks and is apparently not yet available to the public?'

Few members of the general public, then, would have seen a Stag before the Earls Court Show opened, on 14th October. The Triumph stand was located at the back of the exhibition hall, flanked by the Rover, Rolls-Royce, Ford and Vauxhall exhibits. One Stag sat on a raised turntable, and there was another in the main area of the stand; and, for those who could not get close enough, there were two more Stags on the Standard-Triumph (Liverpool) coachwork stand just on the other side of the Ford stand.

Production of the Stag was troubled by any number of hold-ups in these early years and Triumph never succeeded in matching production to demand. As a result, there were long waiting-lists throughout the period of the Mark I cars' production, and the lengthy delivery delays undoubtedly discouraged many potential customers from even placing an order. Overseas, where customers were rather less phlegmatic than the long-suffering British about manufacturers' shortcomings, delivery delays caused many orders to be cancelled. This situation was clearly reflected in the production figures for the 1972 model-year (see Appendix C), which were rather less than half those for the previous season when they should have been considerably boosted by the car's launch on the US market.

Running changes

Over the next two and a half years, Triumph made a number of small changes to the Stag's specification. Some of these were dictated by problems encountered in use, and others resulted from feedback about customer preferences; many were also dictated by the need to minimise production costs.

The very first change, if such it really was, occurred in March 1970 before production had really got under way. The first pre-production car, LD 1, had been built with a bright metal trim strip on the sills just below the doors, but it would be the only early Stag to have this feature. From LD 4 (and possibly from LD 3; LD 2 was a US-specification car with full cover plates on its sills), Stag sills were plain. It is hard not to see the influence of BL's cost accountants in this deletion.

The process of building a new car in quantity for the first time almost invariably reveals unexpected assembly difficulties, and the start-up of Stag production was no exception. On the first 20 or so cars, built in March and April 1970, the seams between the rear wings and tonneau panel were leaded-in, but on later cars they were left exposed — probably, recalls John Lloyd, because of Trade Union pressure to stop assembly workers handling lead on health and safety grounds. Over the summer of that year, there was also a change in the engine assembly procedures: the original method of aligning two dowels on the flywheel with two holes in the crankshaft and then inserting four bolts

LD 1 again, this time with its hard top in place during the photographic sessions in southern Europe before the launch.

Above: The very first cars had blue-painted rocker-boxes. This one is on LD 20, registered as XPW 313 H, and actually the first car ever to be sold to a member of the public.

proved tricky, and so the dowels were replaced with two more bolts after 272 engines had been built.

Those cars built before the autumn of 1970 had only Triumph badging on the outside, although Leyland had stamped its identity on the interior by means of a large letter L on the rubber pads fitted to clutch, brake and accelerator pedals. However, British Leyland appears to have decreed that all 1971-model Triumphs should have small blue-and-white badges with the Leyland wheel symbol on the lower edge of each front wing, behind the wheelarch. TR6s, Spitfires and GT6s, saloons, and Stags were all given these badges — and photographs show that LD 1 (RRW 97 H) was also fitted with them retrospectively, although the RVC-registered press demonstrators were not.

However, for some reason (probably cost-saving), these badges were not fitted in production but were supplied with the cars for dealers to fit. Some cars were sold without the badges, either because Triumph failed to supply them or because the dealer forgot to fit them, or possibly because a customer specifically asked for them to be left off. This has contributed to endless controversy about whether or not they are 'correct' on early Stags;

Opposite page
The first Stag to be exported to Italy was loaned to Michelotti for a time. Here it is at his premises, complete with Italian-specification front number plate box and price sticker (in Lire) on the windscreen. The grille badge, like those on the rear wings, had a grey background on Mark I cars.

Above: Stags built before January 1972 (except those for the US market) had this type of air filter box, with twin 'trumpet' intakes. Note the inverted Triumph name on the rocker-box. Some early cars — like the one pictured on page 46 — were fitted with one which had the casting the right way up, but it appears that a batch was made with the name inverted and that Triumph decided to use them in production!

Opposite page
Top: Michelotti's Stag again, this time showing the special rear number plate plinth used on Italian-market cars. Like all the first-sanction Stags, this one had no trim strip on the sills. Note the rear quarter-windows, standard on all Mark I soft tops.
Below: A first-sanction Stag in side view. The lettering on the Michelin XAS tyres was not normally picked out in white, as it has been here.

probably the nearest it is possible to get to a definitive statement on the issue is that Triumph's masters at British Leyland *intended* the badges to be fitted to every car.

Around 500 Stags had been built by the time the next two changes were made. Once again, one was intended to streamline production and save costs, and the other was intended to improve customer appeal. Thus, from about October 1970 on the assembly lines, a stainless steel fillet fitted at the top of the doors' closing faces was deleted, and grained upholstery material replaced the un-patterned type seen on the earlier cars.

Modified front suspension struts arrived some four months later, and the first car to be fitted with them was LD 921. More changes came in March 1971, after around 1,000 Stags had been built, when improved hood catches were fitted and the bonnet release was repositioned on the left-hand-side. This modification was mildly irritating to UK owners, used to having all the controls logically grouped together on the same side of the car, but there was a certain logic to it from Triumph's point of view. It was important to sell Stags in the USA; assembly of US Federal-specification Stags was just about to begin for a summer 1971 launch; and

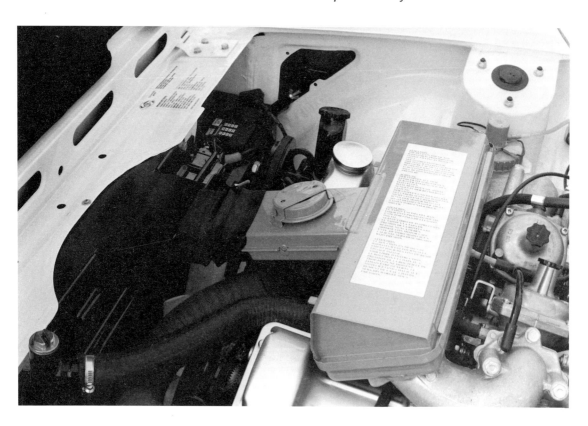

Opposite page
Top: *The second-sanction cars, built from January 1972, had the Federal type of carburettor air intake with a thermostatically-controlled valve, which is clearly visible here as the circular appendage on top of the inlet duct.*
Bottom: *Stag body production at the Speke plant in Liverpool.*

Right: *Stag bodies at Speke receiving final attention before painting and trimming.*
Below: *Inspection of a Stag body at the Speke plant.*

Left: On Mark I Stags (except Federal and Australian models), there was a plain cushion at the top of the backrest, and no provision was made for fitting head restraints.

US customers had complained bitterly about left-hand-drive cars with bonnet releases on the right-hand side. And so, to keep the US market happy and to save costs by simplifying assembly procedures, the bonnet release on all cars was fitted in the US market position.

A change made to the fuel tank at about this time had also been brought about by development for the US market. One of the requirements of the Federal safety regulations was that fuel should not be spilled during a collision, and to that end Triumph had given the Stag a 14-gallon tank with its filler tubes installed in such a way that only 12¾ gallons could be accommodated: the remaining space was expected to allow for fuel displacement during a rear-end collision. However, further development showed that a separate overflow tank was a more reliable way of achieving this end, and so the filler tubes in the main tank were repositioned to allow a full 14-gallon fill. This had the advantage of increasing the Stag's touring range, and also looked better in the specification!

March 1971 also saw a fourth change, and one which, in retrospect, has a certain prophetic irony about it. Early Stags had been plagued with a faulty coolant temperature warning light, which tended to come on for no good reason. Quite what the cause was is unclear, and Triumph's own service engineers were unable to sort it out. In a bulletin sent out to all Triumph dealers, the company recommended a brutally simple 'cure' — disconnect the warning light! And from March 1971, the cars were built with the warning light disconnected. In view of the fact that the temperature gauge was normally masked by the driver's left hand, and in view of the Stag's later reputation for terminal overheating, this deletion was just one more in a series of decisions which did the car no good at all.

It is not clear how many Stags were authorised by Triumph management under the first build sanction. Perhaps the figure had been 4,000. In any case, the last car actually built in that first sanction appears to have been LD 3900, which came off the lines on 9th November 1971. The second-sanction cars, numbered in a new sequence starting with LD 10001, had in fact started coming off the lines on 4th November, or possibly slightly earlier; and some early US-market cars were built in the sequence LE 7500-9000, although no production records for these have been found.

The second-sanction cars incorporated further small modifications. The oil filler cap on the right-hand rocker box cover now had two fixing pins instead of three, there were black plastic glovebox hinges instead of the earlier chromed type, and the safety belt buckles on the transmission tunnel were now in two separate blocks instead of a single block. Three months later, probably from car number LD 10747, a stainless steel trim strip was added to the sills, exactly like the one always fitted to LD 1: presumably customer requests for something to slim the Stag's side elevation had prevailed over the cost accountants' wishes this time!

Cooling system problems were also on the

Above: The Stag's concave dashboard was very neat, with a light wood veneer.

Below: Characteristic of Mark I models were the black instrument bezels, downward-pointing needles on the minor gauges, and rather fussy markings on the two larger dials.

Left: Multi-function warning light dials were a Triumph characteristic. On the Mark I Stag, there was a segment for the temperature warning light (visible on the left) and a single warning light (centre top) for the indicators.

Below: Mark I Stags had five-spoke wheel covers as standard, with the Leyland wheel symbol in the centre and blacked-out panels between the spokes. First seen on Triumph models in 1968, but then without the British Leyland logo in the centre, these trims helped give the Stag a family resemblance to other Triumphs.

agenda again, and the cars built from January 1972 (again beginning with LD 10747) had a sealed cooling system running at 20psi with a new radiator and expansion tank. In addition, there was a new water pump, with a 12-vane impeller instead of the earlier six-vane type, and a simplified by-pass hose. The sealed cooling system should have prevented the water loss which led to overheating when the high-mounted water pump became starved of coolant if only that system had remained leak-free. Unfortunately, it did not, and cooling system troubles would continue to cause Triumph headaches for the rest of the Stag's production life.

As well as attending to cooling system troubles in January 1972, Triumph streamlined production by standardising the air filter box already in use on US Federal cars. This had a duct leading to the radiator instead of the twin 'trumpet' intakes of the original home-market type, and its purpose was to provide the engine with warm air from behind the radiator when needed. It achieved this by means of a temperature-sensitive inlet control valve, a remarkably simple piece of equipment which had actually been developed by the Austin-Morris division of British Leyland and would later be used on other Leyland cars, such as the Rover 2200TC models.

Second-sanction cars ended with LD 14158 on 10th November 1972 and, as before, assembly of cars under the new sanction had already begun. Third-sanction cars began with LD 20001, and started coming off the lines on 17th October 1972.

Above and next page: *Triumph was keen to promote the Stag throughout the whole of Europe, not just in the United Kingdom. This one has Austrian plates and is probably one of the earliest left-hand-drive examples built. (MRP)*

Confusingly, these '1973 models' did not differ from their predecessors, although overdrive was now made standard on Stags with manual transmission. A total of 1,230 cars would be built before further major changes were introduced in February 1973. Those revised cars, the Mark II Stags, will be discussed in Chapter 5.

The early Stags in service

By the time the Mark II Triumph Stags were announced, word was beginning to spread of the Stag's cooling system troubles. Triumph already knew that poor manufacturing quality control was the main cause, but Leyland management seems to have been powerless to rectify the situation. To have acknowledged in public that there were manufacturing problems would have been to invite

trouble with the volatile Trade Unions, and so it was in Leyland's best interests to play down any stories of cooling system problems on which the press invited them to comment.

As a result, the press did not report on the fast-growing epidemic. The Triumph press office could point to genuine cases where cooling system troubles had developed because an owner had failed to observe the maintenance schedules recommended in the handbook for the car, and the motoring press would undoubtedly have been sympathetic. *Autocar* encountered no cooling system trouble at all with its long-term test Stag during this period, although there can be little doubt that the professional attention which the car received contributed to this unblemished record. There was also an element of luck in it: the Stag which *Motor* had on long-term test two years later did have serious engine trouble which was caused by a manufacturing fault.

Only *Motoring Which?* for April 1973 hinted that all might not be well with the Stag, but it was annoyingly imprecise about the car's failings. Comparing its own test car and readers' Stags with a similar sample of Audi 100 coupés and Rover 3500S saloons in a report on fast four-seater cars, it concluded that the Stag was both too noisy and had a poor reliability record. The magazine also commented that the condition of Stags on delivery was very poor — a dreadful indictment of the inadequate quality control standards and dealers' pre-delivery inspection checks which were then becoming the norm throughout the British Leyland empire.

CHAPTER FOUR

The Federal Stags: 1971-1973

The United States of America was Triumph's biggest export market in the 1960s, so it was inevitable that US-market considerations should have conditioned much of Triumph's thinking on the Stag. Sales performance there was implicitly understood as being a key element in the model's viability and it would probably be true to say that, but for the failings of British Leyland management in the mid-1970s, the Stag would have been killed off completely when it failed in America.

Right from the beginning, senior Triumph management was concerned about the car's potential as a dollar-earner. When Harry Webster presented his formal product proposal to the Triumph Board in July 1966, it contained a keen price estimate of $3,875 in New York which would make it cheaper than every competitive model except (as the proposal noted) the Austin Healey 3000 Mark III and the Sunbeam Tiger. At that same meeting, Triumph's Sales Director Lyndon Mills was optimistic about US market sales. His department believed the Stag would sell well in the USA, where the appeal of the luxury sports car was now increasing at the expense of the small sports car. From the comfort of the Boardroom, therefore, the Stag looked like a good bet for the American market.

However, in developing the car for that market, Triumph would be faced with a whole set of new problems. The first was a marketing problem: the car was a new departure for the company and it would have to decide how best to break into the market for luxury sports cars which had been identified at that July 1966 meeting. The second and third were engineering problems: Triumph would have to design the car to meet the forthcoming Federal exhaust emissions standards and would have to ensure that the car's structure complied with the new Federal safety standards. Chief Engineer John Lloyd was sure that his department was equal to these challenges, but there was no doubt that their resolution demanded a great deal of Triumph's small engineering staff.

So what exactly was the market for luxury sports cars? As yet, it had not become clearly defined, although there can be little doubt that the trendsetters had been the American domestic 'personal cars' like the Ford Mustang and European imports like the Mercedes-Benz SL: two cars sufficiently different in both concept and price to leave

Above: The US-market styling development car was X 790, the sixth Triumph-built prototype. The UK number-plate is strapped rather incongruously across the US-specification licence-plate holder and a racing-style door mirror is being tried out. This picture dates from March 1970.

interpretation of the market trends wide open! However, it would probably be true to say that the luxury sports car was essentially a 2+2 rather than a pure two-seater, and that it provided good weather protection instead of the sidescreens and ill-fitting convertible top which had characterised the pure sports car market.

The new Federal legislation had first been announced in 1966 — the very year in which the Stag product proposal was approved. It had been drawn up primarily in response to the demands of domestic pressure groups. These groups had been inspired by Ralph Nader's attack on safety standards in domestic cars (notably the Chevrolet Corvair, which was severely criticised in Nader's book, *Unsafe At Any Speed*), and by the growing 'smog problem' in major US cities, which was thought to be caused largely by vehicle exhausts.

Exhaust emissions legislation, which would limit the amount of noxious gases permissible in vehicle exhausts, was due to affect all vehicles sold in the USA during the 1968 model-year, while the Federal Safety Standards would be enforced for 1969 models. These standards covered the ability of a car's structure to withstand various types of collision without harm to the occupants, and included regulations about protrusions inside the car which could cause injury if its occupants were thrown about in a crash.

Development

These, then, were to be important factors in developing the Stag and in bringing it to the US market. But Triumph decided that they should concentrate on getting the basic shape of the vehicle right as a priority, and adapt the Stag to meet US market requirements later. Of course, one eye was kept on eventual US needs right from the start, and as early as May 1967 the Triumph Board learned that the designers were having to change some features of the car as designed by Michelotti because they would not be suitable for the US market. It is also true that the second Triumph-built prototype, put on the road in March 1968, was used to develop certain items for the US versions of the

Above: *The second car off the production line was built up as a US-market model and sent out to Triumph's importers in New Jersey. It is seen here wearing Illinois number-plates which, in publicity pictures, were covered over by large plates bearing the 'Stag' name. Note the body-colour door mirrors.*

Right and below: *External features of the Federal Stag — wire wheels, stainless steel sill panels, front and rear side marker lights — are clearly apparent on this well-maintained example. (Heathrow Stag Centre)*

car. But the real development work on the US-market Stag did not begin until the end of 1969. By this time, the first pilot-production European car (LD 1) had already been built.

The car on which the full US specification was developed was X 790, which was built up in the autumn of 1969 and registered as RHP 659H in September or October of that year. It was, of course, built with left-hand drive, but it did not have all the eventual US market features from the beginning. These were developed individually over the next five or six months and fitted to the prototype car as and when they were ready.

Thus another prototype car, X 802, was subjected to the barrier crash test demanded by the new Federal Safety Standards at MIRA in December 1969. The same car underwent further crash testing in April 1970, then again in January 1971 and March 1971. The emissions-controlled engine seems to have been ready by August 1970, when it was photographed for the Triumph archives. RHP 659H itself seems to have been used mainly for styling development work at first, and then went out to the USA in spring 1970 for air conditioning and cooling systems tests at Pikes Peak. Later that year, it was fitted with a prototype fuel-injected engine which was probably being evaluated for eventual US market use.

While all this was going on, Triumph shipped a car out to the USA for their dealers there to examine. Although not to full US specification, it certainly seems to have had most of the eventual US-market styling features. The car was LD 2, the second hand-built pre-production vehicle, which was built up in February 1970 and went out to the headquarters of British Leyland Motors, Inc in Leonia, New Jersey two months later, sailing on the *Atlantic Crown* on 13th April. The car was finished in Saffron with a black interior, and had automatic transmission. It also had wire wheels, air conditioning, stainless steel sill covers and side marker lamps, just as production cars would have; and it is this car which appears in so many early US-market publicity pictures of the Stag.

Apart from further barrier-crash tests, most of the next year seems to have been taken up with preparing the way for the US launch. The Stag was announced in July 1971 for the 1972 model-year, although Spen King maintains to this day that he was opposed to a US launch at this stage. 'The car simply wasn't ready for the American market,' he says, 'but of course it had to go there because that's where it had been designed to go.' His misgivings were to prove well-founded...

The specification of the US Stags closely parallelled that of the European cars, although there were a number of important differences. Thus, the sills were covered with stainless steel panels and, to meet the latest Federal regulations, there were side marker lights front and rear, with a discreet 'Stag' plate badge below the rear one. The front seats had integral head restraints (which took the place of the top cushion on seats for other markets), again designed to meet Federal safety regulations, and the front and rear number plate boxes were designed to accept US-style number plates. Lastly, centre-lock wire wheels (long a favourite on TR sports cars for the US) were fitted as standard.

Probably the most difficult part of the US model Stag's development had been getting the engine right. As many makers found to their cost, exhaust emissions control gear tended to have the effect of smothering the engine and so reducing power and increasing fuel consumption. Thus, the compromise which Triumph had reached for the Federal Stag was not a bad one. With a compression ratio lowered slightly from the 8.8:1 of European engines to 8:1, the Federal engine gave 127bhp at 6,000rpm and 142 lbs/ft of torque at 3,200rpm. With 18bhp and 38 lbs/ft less than its European counterpart, the Federal Stag was thus at a considerable disadvantage on paper, although in fact it was capable of reaching the same maximum speed. The losses were in acceleration, both in the gears and from a standing start, as the figures in Appendix E make clear. Fortunately, it had lost none of the refinement and smoothness of the European version, which was quite an achievement: many manufacturers found that their early emissions-controlled engines were much rougher than their pre-emissions control counterparts. And, as far as fuel consumption was concerned, losses were minimised by means of the thermostatically-controlled hot air intake which became standard on Stags for other markets a few months after the first Federal cars had been built.

Stag.
A new kind of Triumph.

For years, Triumph has been making fine sports cars for people who love cars. Now Triumph introduces a sports car which loves in return.

To all the things that make a true sports car good to drive, the Stag adds all the things that make a car good to ride in. Stag adds power to Triumph's rack and pinion steering and front disc brakes. It combines more room and comfort with Triumph's road hugging, independent suspension.

The engine is a big, smooth V-8, making the Stag the fastest car in the line. There's true 2 plus 2 seating, electric windows and a solid, padded roll bar, even when the top is down.

Big, wide doors make it easy to get in and out, controls are in easy reach of your fingertips, and both the reclining bucket seats and the padded steering wheel are fully adjustable.

The Stag is a new kind of Triumph, a powerful over-the-road car built by the biggest maker of sports cars in the world.

Base price is $5525,* including chrome wire wheels and radial ply tires.

Options include a detachable hard top with a heated rear window, automatic transmission, air conditioning, and the pleasure of test driving the Stag at your nearest Triumph dealer. **Stag**

*Manufacturer's suggested retail price, P.O.E., optional equipment, destination charges, dealer preparation charges, state and local taxes, if any, not included. British Leyland Motors, Inc., Leonia, N.J. 07605.

Above: American sales literature sang the praises of the Stag in characteristically brash style. Price of $5,525 (or $5,575 through West Coast dealers) did not include a number of virtually essential options, which would add a further $1,000 or so and make the car a less attractive purchase than it might at first have appeared.

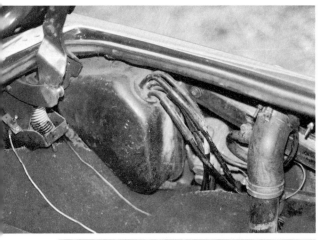

Below: The thermostatically-controlled air intake was fitted to all Federal Stags from the beginning, as this 1971 model shows. It was fitted to cars for other markets as well from January 1972.

The US launch

The basic price of the four-speed manual Stag was announced as $5,525 on the East Coast of the United States, a figure which was inflated to $5,575 on the West Coast because of the additional shipping costs from the port of entry. The hard top, originally announced as a $248 option, was then included in the standard specification and the East Coast price became $5,650 by October 1971. However, options could easily add another $1,000 to these prices. An AM/FM radio cost $135-$145, air conditioning $495, overdrive $175 and the automatic transmission $219. Later in the 1972 model-year, leather upholstery would become a further extra-cost option (although no cars are known to have had it from new). Even though the base price was attractive and the options relatively few in number, Stags for the USA were therefore not quite as cheap as they looked at first sight.

Press reaction to the car was lukewarm, which was surely a blow to Triumph. None of the leading motoring magazines praised the overall concept of the car. *Road and Track* claimed that 'perhaps the Stag will find its greatest potential market among those who want an SL (Mercedes-Benz) but can't afford it', and *Sports Car Graphic* noted that 'it operates smoothly and quietly in the best American non-car tradition, which,' it added somewhat cynically, 'should make it an immediate hit.'

The Federal Stags: 1971-1973

Opposite page
Top: *The early Federal Stags had a fuel expansion tank to meet safety regulations. It was developed on X 783, the fourth Triumph-built prototype, and is seen here on that car. Later Federal cars had a smaller expansion canister in place of the large tank.*
Middle: *US regulations also required marker lights in the rear wings, and these were set off by neat 'Stag' badges which were not used in other markets.*

Above: *Federal Stags had integral head restraints in their front seats. These replaced the plain cushion panel at the top of the seats fitted to cars for other markets.*
Right: *Interiors of the Federal models, as shown in this shot of a 1971 model, were otherwise similar to those for other markets.*
Below: *The hazard warning light installation differed from that fitted to Stags for other markets in 1974. This one is on a 1973 MkII Federal model.*
Below, right: *Rear lap belts with inertia reels were fitted to Federal cars. The reel was con-cealed behind the side trim.*

The Stag did have its good points, though. *Sports Car Graphic* liked the interior design, found the headlamps gave an excellent spread of light and praised the thought that had gone into making the car meet the safety and emissions standards without destroying its performance. *Road and Track* spoke favourably of the handling and the ride quality, and echoed *SCG*'s praise for the interior design, while *Car and Driver* liked the inertia-reel safety belts, the good visibility, and the overdrive. But these were details, and the lists of 'dislikes' were much longer...

Sports Car Graphic thought it was $1,000 too expensive, found the automatic transmission made the car irritatingly gutless, noted that the hard top was cumbersome to install and remove and that the bright trim could be damaged in the process, and objected to the over-enthusiastic low-fuel warning light. Prophetically, the magazine also commented that the Stag's engine overheated too readily under load in hot weather. Overall, it pronounced that 'the four-speed version of the Stag is most acceptable, but we really can't think of many good things to say about the automatic.'

Road and Track singled out the power steering, which it claimed lacked feel and was noisy. It also criticised the brakes, in particular rear wheel lock-up in a panic stop and the handbrake's failure to hold the car on a steep slope. Both hard and soft tops fitted badly, created wind noise and leaked, and, in general, 'true quality of assembly is not one of the Stag's outstanding features.' The magazine concluded, 'as it stands it has too many detracting irritations to be a really satisfying car even after its basic character is accepted.'

Car and Driver also disliked the Stag's 'insensitive' power steering and its premature rear brake lock-up. It found the manual gearbox notchy and thought that the ride/handling compromise gave less than ideal handling. The Stag, it declared, had 'the same feeling of rough-hewn capability that has been a trademark of all Triumph cars sold in this country' — a most unfortunate verdict on a car which was supposed to be a refined grand tourer!

Right from the start, then, things did not go well for the Stag in the USA. Worse was yet to come. The engine problems which afflicted European cars showed up even more readily in the Federal models, partly perhaps because of the harder use to which American drivers typically put their cars, and partly perhaps because the emissions-control modifications made the V8 even more sensitive than it was in European form. One way or another, major warranty work relating to the engine pretty well wiped out any profit Triumph might have made from the Stag in its first year of sale in the USA. News of this soon got around, and the lukewarm reception in the motoring press did the rest: Triumph had been hoping for a modest 2,000 Stag sales in the US market in the 1972 season, but in fact they sold well under half that number.

The 1973 Federal models

Meanwhile, the Triumph engineers were working on improvements for the 1973 model cars. Many of the changes would parallel those for the European 'Mark II' models, and Triumph photographic records suggest that the 1973 Federal specification had been settled by May that year, ready for a July introduction. There were 'Mark II' seats with adjustable head restraints, coachlines on the sides, five-spoke alloy wheels in place of the wires and a cowled door mirror instead of the earlier flat type. Overdrive was standardised with manual transmission, and there had been modifications to the engine, mainly to cope with the stricter exhaust emissions standards which would be enforced for 1973 cars.

The 1973 Federal engines had reshaped combustion chambers and an even lower compression ratio, in contrast to the European engines which had an even higher compression ratio than before. Advertised power, however, remained the same, with the 127bhp now available 500rpm lower down the rev range. Torque was up slightly to 148 lbs/ft, though the peak was now 300rpm further up the rev band. It all added up to a reasonable compromise, which at least did not leave the Federal Stag with less performance than before, even though it could ideally have done with a signficant performance boost if it was to improve its reputation with US buyers.

That was not to be, however. The leading motoring magazines showed no interest in the 1973 revisions, and it is doubtful whether many potential buyers would have been aware of them unless they had visited a Triumph showroom. Stag sales did increase in the USA during 1973, although they still did not meet the total originally anticipated for 1972. And still, Triumph was faced with a continuous stream of engine problems on their Federal cars. Triumph historian Graham Robson has estimated that 75 percent of all Stags sent to the US in the 1972 and 1973 seasons suffered valve-gear problems.

This was a nightmare which could not be allowed to continue. The Stag's bad reputation was beginning to blow back on to other Triumph models, and warranty work was causing the company to make a loss on every Stag it sold in the USA. Worse, more development work would be needed to keep the car saleable, as the Federal emissions regulations were due to be tightened up. On top of that, California — where the smog problem was greatest — had decided to go its own way after 1973 and to demand even stricter control of exhaust emissions. If the Stag was to remain saleable right across the USA, then, Triumph would need to develop two new versions of the engine.

It was all too much. Triumph management decided to withdraw the car from the US market at the end of the 1973 season, and the Stag ceased to be a catalogued model for 1974. A few of the 2,871 cars built in two seasons of production remained unsold as late as 1975. The Stag had failed in the one market which could have been the key to its success, doomed by a lack of clear identity, appalling quality control, and the difficulty of keeping pace with tightening exhaust emissions regulations.

Below: *The US Stag sales brochure included this V8 engine cutaway drawing, pronouncing it 'an engine along racing lines, tamed for road use in the Triumph Stag'.*

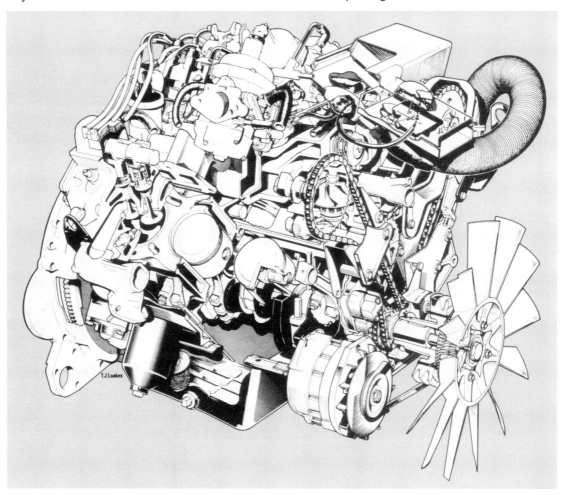

Californian Dave Bergquist writes frequently about Stags both in America and for the UK's Owners Club magazine. His car is shown here in Arizona (with another export, London Bridge, in the background); on San Francisco's Hyde Street Hill, overlooking the Bay; and on the West coast with Alcatraz Island just discernible in the distance. This is one of the final Federal models with alloy wheels. The bump-strip has been added for protection. (D. Bergquist)

CHAPTER FIVE

The Mark II Stag: 1973-1975

The fact that Triumph did not manage to introduce their Mark II version of the Stag to coincide with the start of the 1973 model-year is probably of little concern to anyone except motoring historians, who like to have things neatly pigeon-holed. As far as Triumph were concerned, the fact that the two did not coincide was probably a result of the production hold-ups which continued to affect the Stag.

When the Mark II Stags were announced in February 1973, it was clear that Triumph had put a great deal of effort into improving the car, for almost every criticism levelled at the original cars had been attended to. The major exception was that the Stag's performance had not really been improved, even though the sales catalogues now quoted its top speed as 120mph instead of 118mph. It was, of course, true that the Stag was already fast enough for the majority of its customers, and Triumph had therefore elected to spend the money available for development on other areas of the car.

There had nevertheless been several engine changes, and all of them for the better. Higher compression pistons with domed tops had been fitted, and the combustion chambers in the cylinder heads had been reshaped, although the extra 1bhp from the revised engine would hardly have justified the expense if the main benefit of the changes had not been to give smoother running. Torque, in fact, was slightly down on the earlier engines, but the difference was not noticeable in use. Quieter running was also a characteristic of these Mark II engines, although the main reason was probably the smaller-bore exhaust tailpipes which came with them. Less noticeably, the new high-compression engines had also been fitted with a new alternator.

The 1973-model TR6 sports car had already exchanged its earlier A-type Laycock overdrive for the newer J-type, and Mark II Stags followed suit. This was purely the result of rationalisation at Triumph, however: from the customer's point of view, there was no discernible difference, as gearing remained the same and the overdrive was still operated by a sliding switch in the gear lever knob. Brakes and steering were unchanged as far as their mechanical components were concerned — but there had been an important change to the steering.

Some road-testers of the Mark I Stag had complained that the power-assisted steering was too vague — that it lacked 'feel'. Triumph had

Above: *The arrival of the Mark II Stag in February 1973 elicited virtually no comment from the motoring press. Various changes had been introduced, though performance was unaltered. (MRP)*

responded to this simply by fitting a smaller-diameter steering wheel, which had the effect of increasing slightly the turning effort required from the driver and of making him feel more in touch with the front wheels. The change followed motor industry trends, too, because smaller-diameter steering wheels were now becoming fashionable.

On the Mark I cars, there had also been nowhere for the driver to rest his left foot, so Triumph had catered for that one by standardising a small foot rest on the transmission tunnel in right-hand-drive cars. And, though he might not have noticed the change very readily, the driver of a Mark II Stag looked out through a laminated windscreen instead of the zone-toughened type fitted to Mark Is (on which the laminated screen was optional).

The driver might well have noticed if the new option of Sundym tinted glass was fitted, however.

Opposite page
Mark II recognition features visible in this picture are the blacked-out sills, the twin coachline and the unpainted panels on the wheel-trims.

Many Mark II Stag customers probably specified this simply because it looked good, but it did bring very real benefits in cutting down heat penetration into the interior of the car in hot weather and, to a lesser extent, in cutting down glare. In addition, the driver of a Mark II Stag might well have spotted that the design of the windscreen wiper pantograph on his side of the car had changed slightly.

There were also several interior changes. Most obvious of these were the redesigned seats, with provision on the front pair for optional adjustable head restraints and backrest release catches positioned more accessibly than before. The basket-weave panelling on the front seats now reached all the way up the backrest, and consisted of ten panels instead of the nine seen on Mark I seats. In addition, the upholstery material was now a flameproof vinyl, in the interests of greater safety, and a new dark brown colour known as Chestnut became available with certain exterior finishes.

Compared with the Mark I cars, there were many other interior changes, although most would only have been apparent to someone with a thorough knowledge of the superseded models. The instrument bezels were now chromed instead of black, the needles of the smaller gauges pointed upwards instead of downwards (the same change had been made on 1973-season TR6s), and the speedometer and rev counter had clearer numerals, without the rather fussy-looking white band beneath them. At night, brighter instrument illumination was available, this time with the green tint increasingly recognised as more restful than the pure white light seen on Mark I Stag instruments. The stalk switches now offered an intermittent mode for the windscreen wipers, but the parking lights facility had been deleted.

On the console, the cigar lighter had changed, too. The carpets were now of different material, and new Kangol safety belts were fitted. There was a single courtesy light in the centre of the roll-over bar instead of one on each B-pillar, and improved door locks completed the interior package.

Outside, the changes were less extensive, but their cumulative effect was altogether greater. Mark II Stags were immediately recognisable by their matt black sills and tail panels, by the twin coachlines down their flanks (which were stuck on rather than painted), and by black instead of grey backgrounds to the 'Stag' badges. Two new paint colours had also been added to the range. At the rear, the number plate lamps had also been moved from the plinth on the rear bumper (which was now smaller) and fitted into the underside of the boot lid. Less visibly, there were stronger bumper mounting brackets. And the Michelin XAS asymmetical-tread tyres fitted to Mark Is as original equipment were now supplemented by Avon

Left: From behind, the Mark II Stags could be recognised by their blacked-out tail panels and, when the soft top was erected, by the absence of quarter-windows. This car also has the remarkably attractive five-spoke alloy wheels which were introduced as options for the Mark IIs.

radials of the same size. Cars left Canley fitted with either, depending presumably on which manufacturer had delivered the last batch of tyres.

Most of these changes made the Stag look more distinguished, but the decision to delete the black panels from the standard wheel trims had rather the opposite effect. John Lloyd agrees that the unpainted panels had a cheapening effect on the car's appearance, but adds that the change was made purely to help the Mark II models look different from Mark I cars. There was compensation, however, in the shape of the attractive new five-spoke alloy wheels which had already been seen on 1973-model Federal Stags. These wheels — an extra-cost option — really did look the part. So attractive were they, in fact, that they would later appear also on Triumph's 2500S saloons and estates and, in one of British Leyland's attempts to commonise components between marques, on the final MGBs.

Lastly, a marketing policy change meant that Mark II cars came either as pure soft tops or as 'hard top' models with the soft top as well; Mark Is had come as one or the other, unless both tops had been ordered at extra cost. The actual soft top on Mark IIs also differed. On Mark Is, the rear quarter-windows had tended to get marked or torn by catching in the hood mechanism. The simplest way of dealing with this problem had proved to be to delete the windows altogether, and so Mark II Stags had plain rear quarter-panels in their soft tops, which did nothing for rearward visibility. Coverings were now Mercedes mohair instead of double-duck, and were still available only in black although a beige headlining replaced the black one of the Mark I 'three-window' soft tops.

Press reactions to the Mark II

It is one of the unexplained mysteries of Stag history that no major motoring magazine published a road test of the Mark II car when it was announced. There was certainly no shortage of press demonstrators, and surviving records show that, by March 1973, there were four Stags on the Triumph press fleet, registered FDU 735 L and FDU 751 L to 753 L. Yet it was autumn and Motor Show time before any road tests appeared in print; and there were precious few of these. By then, many of the magazines probably considered a car which had already been on sale for eight months too old hat to be worth writing about.

Motor very much appreciated the Mark II revisions. 'Detail improvements have bettered an already good car,' was the way it summarised its test of FDU 735 L, published in the 27th October 1973 issue. 'This test of the updated Stag reaffirms our initial impression that the car is not only unique in character and a highly desirable property, but that the standard of finish makes it a world-beater at the price.'

Motor Sport's Bill Boddy was less impressed,

however, reporting on FDU 752 L in the magazine's September 1973 issue. 'My initial impressions of the Stag were as unfortunate as before,' he wrote, going on to complain about the light power steering and what he called 'the weaving effect of what seemed to be too-hard independent rear suspension'. He concluded that it was still not a sporting car, but that it was 'a nice motor-car' and quiet enough to be 'a veritable Rolls-Royce among sporting cars.'

Motor also ran a 1973 Stag on its long-term test fleet. TMM 964 M actually belonged to the magazine's Publishing Director, John French, and he reported on progress from time to time in the magazine. After 10,000 miles, his report published in *Motor* dated 21st September 1974 revealed that the car had been a thoroughly enjoyable means of transport, apart from a freak occurrence when a battery lead had caused a small underbonnet fire after chafing on a power steering pipe and on an oil pipe. Once, the report noted, the engine had got a little hot; but this was put down to the effects of a long, fast run while using 3-star (97-octane) fuel instead of the prescribed 4-star (98-octane) type.

That incident of overheating was in fact the beginning of a major problem, of the type with which too many Stag customers were already familiar. By the time of the 20,000-mile report on TMM 964 M, published in *Motor* dated 28th June 1975, the engine had been well and truly wrecked. A 'slight casting obstruction' in one of the engine's waterways had caused that first bout of overheating, and it caused it again after a run of 'approximately 100 miles on the Autoroute du Sud at between 90-100 miles per hour.' This time,

Above: This close-up shows the black background to the badges which came in with Mark II models; Mark I Stags had grey backgrounds. The coachline, with its narrow upper and wide lower bands, is also visible.
Below: Instead of twin courtesy lamps, Mark II Stags had a single lamp in the centre of the roll-over bar.

Right: Seats on the Mark II also came with adjustable head restraints.

Above: Mark IIs had a smaller plinth on the rear bumper, and this no longer contained the number-plate lamps...

Below:. ... which were now recessed into the overhanging lip of the boot-lid.

perhaps accentuated by a minor timing maladjustment, the engine oil also overheated and burst out through the oil filter, leading to the demise of two big ends and to contact between two valves and the pistons they served. The car was examined by Triumph engineers, who discovered the casting flaw, and the magazine made no comment to suggest that other Stag owners had experienced similar problems. But the Triumph engineers, at least, must have known...

Autocar also took a Stag on its long-term test fleet, this one a slightly later car registered WYH 514 N. Maurice Smith was its primary driver, and he reported on 24,000 miles with the car in the issue dated 12th February 1977. In two and a half years of use, the car had given no trouble of note except for a fuel pump failure. There had been some minor electrical problems, too, but the car had proved both reliable and enjoyable transport. Smith did criticise certain details, however. 'I would change the design and position of the segmented warning lamps dial which I find ugly and usually obscured by the driver's left hand,' he wrote. 'The upper seat belt attachments (or guides) are dreadful. They cause the straps to rub the seat back; they tangle and jam and seldom take up the tension.

'The rear head room in the hard top is too limited and if even another inch could be found it would be valuable. The soft top could do with transparent quarter panels for better rear vision. The styling strips along the side of the Mk.2 car should be replaced by proper rubbing strips with rubber inlays. A second filler cap on the port side would often be convenient. The plastic clips intended to hold down the floor panel of the boot are miserable fragile little objects which should be improved.

'I would expect that with some better dampers the initial harshness, often felt when the car is rolling over coarse road surfaces, could now be improved. This is not to be regarded as a serious shortcoming and the ride in general is very comfortable and quiet.'

The Mark II Stags on sale

Stag sales on the home market for the 1973 season were some nine percent lower than they had been

Right: Mark II models built during 1973 had a dashboard which was broadly similar to that on Mark Is, but all the instruments now had chromed bezels, the pointers of the minor instruments faced upwards, and the numbers on the main dials were clearer. On cars built after January 1974, the handbrake warning lamp was repositioned, and a hazard warning switch and seat belt lamp were added.

for 1972, and there can be little doubt that a few reviews in the motoring press when the Mark II cars were announced would have boosted sales and so prevented this downturn. Nevertheless, there appears to have been no shortage of orders, and delivery times for Stags were being quoted as between three and six months (*Autocar*, 13th September 1973) and one year (*Motor*, 27th October 1973). Triumph probably could have sold more cars on the home market if only they had been able to make enough.

The reason for these delivery delays on the home market was probably that production had been deliberately geared to meet export demand. Sales figures for the 1973 season show an overall increase of 22 percent as compared to 1972, and export sales had improved by more than 130 percent. This was an astonishingly encouraging improvement, especially at a time when sales were falling in the US market, and it was mostly attributable to the Stag's launch in Australia.

Exports to that country began in October or November 1972, just as the final Mark I cars were being made, and the first Australian cars had a specification which included US-market style head restraints and 8.8:1 compression engines. However, the majority of Stags sold in Australia must have been Mark II models, as these came on-stream shortly after the car's announcement over there.

Records show that Australia rapidly became the car's best export market.

Nevertheless, the successes of the 1973 model-year were to be short-lived. Just as the 1974 season began in the autumn of 1973, war broke out between Israel and an alliance of Arab nations. The conflict was brief, with victory going to the Israelis; but in response, the Arab oil-producing nations quadrupled the price of crude oil and embargoed supplies to those nations they claimed had supported Israel during the war. The result of this was a serious downturn in large-car sales in all the Stag's major markets, as buyers scrambled to buy small, fuel-efficient cars. Stag production was cut back to meet this reduced demand, and the figures show just how serious the problem was: in the 1974 model-year, overall Stag production fell by nearly 40 percent. In the following year, even though oil prices had begun to stabilise and normal supplies had resumed, demand remained low and production fell by a further 16 percent. Worst hit were the manual-transmission cars, a fact which is hard to explain in the light of their better fuel economy.

The Stag was in serious trouble. Sales had tumbled heavily after the 1973 oil crisis; news of persistent engine problems was getting around; and its manufacturers were also fast running out of money. It was in 1974 that the British Government

Above: Production of the Mark II was deliberately geared to meet export demand. Domestic sales fell as a result, but overseas sales rose substantially, helped to a great extent by the Stag's introduction into Australia. (MRP)

asked Lord Ryder to investigate British Leyland's huge losses, and a year later the corporation was nationalised in order to prevent job losses on a massive scale. In these circumstances, there was no money to spend on gearing the Stag better to what remained of its market.

Running changes

Triumph's commission numbers at this stage in the Stag's history are confusing. If the Parts Catalogue is to be believed, the first Mark II Stag was numbered LD 20001, but there are some problems with accepting this to be the case. Firstly, car number LD 20001 was built on 17th October 1972, four whole months before the Mark II cars were announced; and secondly, all Mark II cars with overdrive are supposed to have had the J-type overdrive in place of the earlier A-type, and yet the first car with a J-type overdrive seems to have been LD 21230, which was built some time around February or March 1973.

Whatever the truth behind that, it is clear that very few specification changes were made to the cars between the introduction of the Mark II cars and the arrival of further updated Stags in October 1975. In January 1974 (at car number LD 31153), the hazard warning light system already fitted to left-hand-drive cars was standardised for all markets, together with a warning lamp which lit up when the ignition was on to indicate when a front seat had been occupied but its associated safety belt had not been fastened. So that this new lamp could be fitted in the top centre of the dash, the handbrake warning lamp was moved to its outboard lower corner. The air conditioning installation option — never very popular — was dropped in March 1975, and then there was a new seat belt warning light from car number LD 38324 in July 1975. One exterior colour was changed for 1974, and the 1975 season ushered in several new paint options, which were clearly intended to brighten the car up and freshen its appeal. That, however, was all.

CHAPTER SIX

The final Stags: 1975-1977

With hindsight, the summer of 1975 would have been as good a time as any to admit that the Stag had failed and to take it out of production. The fact that Triumph did not follow this course of action can probably be attributed to two main causes. First, the car still had a powerful protector in the shape of Lord Stokes, by now British Leyland's President, who liked the Stag and continued to view it as the flagship of the Leyland range. Second, it would be several years before the car which was being developed to take its place — the Lynx coupé — would be ready for production and, if too big a gap were left after the Stag's demise, Leyland would lose to rival manufacturers all the customers the Stag had attracted.

Secure in the knowledge that the Stag would continue in production for the time being, Triumph looked around for the most effective ways of giving it the shot in the arm it needed to start selling well again. Working with a very limited development budget, the engineers looked at four different ways forward. The first was to take some of the weight out of the car and thus to improve its fuel consumption and its performance. The second was to improve the performance alone. The third was to minimise production costs by taking the three-litre V8 out of production altogether and replacing it with the Rover 3½-litre V8, which had the additional advantage of being a well-respected and trouble-free engine. And the fourth option was to come up with a limited cosmetic overhaul. In the event, although research was done on all four options, only the fourth and cheapest of them went into production.

The lightweight Stag was a particularly interesting project, and one which promised a great deal. Exactly what Triumph had done to take the weight out of the car is not clear, but the lightweight Stag tested by Chief Experimental Engineer Tony Lee in November 1975 was some 265 kg (586 lb) lighter than the standard car when fitted with an overdrive transmission. This allowed taller final drive gearing of 3.15:1 (the effective gearing in overdrive was 2.59:1), and the fuel savings achieved were astonishing. At a steady 30mph, the lightweight car consumed fuel at the rate of 47.1mpg in overdrive, as compared to 36mpg for the standard car; at a steady 50mph, the figures were 40.1mpg and 33.3mpg respectively; and at a steady 80mph, the figures were 25.9mpg and

Right to the end, BL persisted in promoting the Stag as the car which had vanquished its continental rivals, claiming that 'all over the world, motoring devotees have set their sights (and their hearts) on a Stag'.

The final Stags: 1975-1977

22.9mpg. However, the lightweight Stag did not go beyond the prototype stage. There was simply not enough money to put it into production.

The idea of putting more performance into the Stag was also an attractive one, and would undoubtedly have given the car's image and sales a boost if it had been carried through to production. In 1974-1975, the Triumph engineers looked at two alternative ways of increasing the output of the Stag engine. The first depended on a reprofiled camshaft in what was otherwise a standard production engine, and bench tests saw 210bhp — a very worthwhile improvement over the 145bhp of the production engine. The second added an extra pair of Stromberg carburettors to the otherwise standard engine, and this four-carburettor V8 put out 190bhp on the bench and would rev as high as 7,500rpm. However, it looks as if neither engine ever went into a car. The reasons are unclear, but fear of increased fuel consumption at a time when fuel economy was still a sensitive issue may well have been one of them.

When the Rover V8 had been tried in the Stag during 1967, rivalries between the Rover and Triumph engineering staffs had sealed the project's doom. By 1975, however, things were very different: since 1972, the Rover and Triumph engineering departments had been amalgamated, and the new Triumphs and Rovers then under development were to share a great many major components. One of them was the Rover V8 engine, which would come with either a five-speed manual gearbox or a Borg Warner three-speed automatic. In practice, several projects were cancelled and the only production Triumph ever to use the Rover drivetrain was the ill-fated TR8 sports car. However, the Triumph engineers had been obliged to accept that their future lay with the Rover drivetrain, and so the idea of putting it into the Stag no longer appeared as heretical as it had eight years earlier.

Precise details of what was done are lacking, but the various stories about the project all agree that the engineers fitted either two or three

The final Stags: 1975-1977

Opposite page
For the final Stags, bright sill covers were standardised along with the five-spoke alloy wheels. (Heathrow Stag Centre)

Below: *The final Stags could be recognised from behind by their body-colour tail panels — a reversion to Mark I practice. The fog lamps beneath the bumper of this example were not part of the car's standard equipment.*

* *The black Rover-engined Stag has now come to light. Car number LD 45328 0 was built in February 1977 and joined the Executive Cars fleet at Solihull (then the BL Rover plant). It was originally finished in Inca Yellow and registered as VVC 730 S. After use by several BL executives, it was sold off and acquired the registration 6080 PW, later being re-registered again as VOP 540 S. It was then bought back by BL in November 1978 and fitted with a Triumph TR8 (Rover V8) engine and five-speed gearbox. The respray in black probably dates from the same time. Why BL should have bought the car back and converted it is far from clear, but it raises the intriguing possibility that there were thoughts of reviving the Stag after the collapse of the Lynx project.*
The car's present owner is a Mr Ray Goff, who lives in Nottingham.

development cars with the latest 155bhp version of the Rover V8. Presumably both five-speed manual and three-speed automatic transmissions were tried. One of the cars* (a black one) is said to have been used by a Triumph manager for some time, and one is said to have been sold off by the factory after development was over, and to have had specially-made rocker covers bearing the Triumph name. From the engineering point of view, there is no doubt that the Rover-engined Stag was a viable proposition, but the project did not prosper. Former development engineer Gordon Birtwistle recalls that the need for a new front panel, radiator, subframe and sundry other parts meant that changing to the Rover engine would be expensive, and Leyland decided it had higher priorities for its scarce resources than the resurrection of a model which had already failed by a large margin to live up to initial expectations.

So, in order to keep the Stag alive for a few years longer until its replacement was ready, Triumph chose the fourth and cheapest option. A mildly revamped Stag went into production during September 1975 and was shown to the public in October at the Earls Court Show. The revamped Stag was essentially a modified Mark II, and never had an identifying name of its own.

Left: *1976-model Stag is pictured here with its proud French owner and companion. (Heathrow Stag Centre)*

Opposite page
Top: *This dashboard belongs to a 1977-season Stag, one of the very last cars to be built. There are plastic fillets in the steering-wheel spokes, but the dashboard is otherwise unchanged from that of Mark IIs built after January 1974. Note the positions of the hazard warning switch, the 'Fasten belts' warning lamp (above and between the two large dials), and the handbrake warning lamp (outboard of the driver, below the air vent).*
Bottom: *Australia remained the Stag's best export market until production ended. This picture shows a 1976 Stag registered in the state of Queensland with an appropriate number - STG 76. (John Frost)*

The 1976 model Stags

The October 1975 changes were disappointingly minor, and customers would have had to look very hard to detect those which had been made in the passenger compartment. Most obvious, probably, was the new deep-pile carpet (specified to bring the Stag into line with other Leyland models); the new button for the speedo trip counter reset and the larger handbrake cover would definitely not have attracted attention, however!

The main reason why these cars can be looked upon as a third phase in the Stag's evolution is that they had certain cosmetic differences from earlier models. Immediately apparent were the stainless steel sill covers, exactly like those fitted as standard to the now defunct Federal Stags and in fact probably specified as a way of using up unwanted stocks! The tail panel was now once again in body colour instead of matt black, and the five-spoke alloy wheels were standard, together with tinted glass all round. In view of Leyland's financial situation at the time, it would probably not be unduly cynical to suggest that the black tail panel had been deleted because it saved time and costs during production, and that the alloy wheels and tinted glass had been standardised to streamline production (and thus again reduce costs)!

Certainly, the changes were not enough to attract significant press attention. None of the major motoring magazines ran a piece on the 1976-model Stag. It is doubtful whether many people at the newly-renamed Leyland Cars cared, though: with the Triumph TR7 already launched on the US market and the new Rover 3500 being readied for its July 1976 appearance, they had bigger fish to fry.

Stag sales did revive a little, although it would be overstating the case to suggest that the revised models were the sole cause. Probably equally likely in accounting for the model's improved fortunes was the fact that large car sales were beginning to pick up again everywhere, now that the first shock of the fuel price increases had worn off. Stag exports, however, did not improve. Already minimal in 1975, they were as good as dead during 1976.

The 1977 model Stags

The Triumph TR6 ceased production in July 1976 and its assembly lines closed down. Stags, always built on the TR6 lines, were therefore transferred to the Innsbruck (2000/2500) saloon lines in August. In itself, this had no implications for the specification. As far as Triumph (or, rather, Leyland Cars) were concerned, the 1977-model Stags introduced at the October 1976 Motor Show were not different enough from the 1976 models to justify a new commission number sequence. In fact, however, the 1977-model Stags were in some ways as

The final Stags: 1975-1977

different from the 1976 models as these had been from their 1975 season predecessors.

The main changes were to the automatic models, which had a Borg Warner type 65 gearbox in place of the type 35 fitted since the start of Stag production. The type 65 was lighter than the older gearbox, and was alleged to be smoother in operation, too. However, the change had not been made out of any real attempt to improve the Stag: it was simply that Leyland was going over to the new gearbox as a matter of policy and the Stag had to fall in line. In fact, fitting the new gearbox must have been just one more source of irritation to Leyland's cost-accountants, for it did make necessary a number of small engineering changes, which all cost money: these included a new propshaft, repositioned oil cooler pipes and redesigned front exhaust pipes. Manual cars, meanwhile, continued with the original type of exhaust.

Other changes were really little more than running improvements. Whether the smaller radiator could be counted as an improvement in view of the Stag's well-known overheating problems is questionable, of course! The steering ratio was also altered, for reasons which must have made sense at the time but which are no longer clear. The 1977 Stags also had a new type of windscreen washer bottle, anti-run-on valves on their carburettors, stalk-type central seat belt mountings instead of the tunnel-mounted 'box', and plastic fillets in the cutouts of the steering wheel spokes. These fillets, says Tony Lee, were added as a safety precaution, to prevent necklaces and other items of jewellery catching in the spokes!

None of these changes improved sales; nor were they intended to do so. Stag sales continued at a low level, and production came to an end in June 1977, as scheduled. A few cars remained unsold and languished in dealers' showrooms for some months. In the UK, these attracted the 'S' suffix registrations which were current between August 1977 and the end of July 1978 or, in a very few cases, the 'T' suffix numbers current during the following 12 months. The actual last-of-line car was retained for the Leyland Historic Vehicles collection and was registered as BOL 88V some time later. This car, specially finished in the British Racing Green which had not been available since the end of the 1975 model-year, now belongs to the Heritage Collection.

... and the future

When the Stag went out of production, Leyland had every intention of introducing a replacement model. The replacement, known as the Lynx, was to be announced during 1978, after a decent interval which would allow old-stock Stags to pass through dealers' showrooms. It may also have been Leyland's hope that this interval would help to dissociate the Lynx from any lingering bad impressions which the Stag's well-publicised failings may have left.

The Lynx was in any case a rather different car from the Stag. In essence, it was a long-wheelbase coupé derivative of the TR7 sports car, and was intended to have the Rover V8 engine and five-speed gearbox. As a coupé, however, it did not offer the Stag's open-air option, and it appears that Leyland saw it as more of a sports car and less of a grand tourer than the Stag.

Everything was on schedule for a 1978 launch until 1st November 1977. That was the day on which the workforce at the Speke factory where the Lynx was to replace the Stag on the assembly lines began a full-scale strike. Negotiations proved fruitless, and after four months Leyland's new Chairman, Sir Michael Edwardes, lost patience and closed the Speke factory altogether. The demise of Speke spelled the end for the Lynx, for it was now behind schedule and there was nowhere else to build it.

Elements of the Lynx were seen again briefly in a project known as Broadside, which was a 2+2 T-bar convertible with the Lynx's floorpan and body styling, but this direct attempt to revive the Stag concept progressed no further than a single prototype, which was rather incongruously fitted with a two-litre O-series Austin engine. Today, the Broadside prototype and two Lynx prototypes survive in the Heritage Collection; but the Stag was never replaced.

CHAPTER SEVEN

Post-factory modifications

Over the years, both professional motor engineers and enthusiastic amateur builders have seen in the Stag an ideal basis for conversion. Some conversions were carried out during the car's production life; others were carried out after production had ceased; and the Stag's popularity and high survival rate means that many more personalised conversions are likely to appear in the future.

The 4x4 Stags

During the 1950s and early 1960s, Harry Ferguson Research developed an adventurous new traction control system for passenger cars, based on ideas proposed by racing drivers Tony Rolt and Freddie Dixon. It incorporated this into its own demonstrator prototype, although its ultimate aim was not to build its own cars; rather, it hoped to interest major manufacturers in the system and to supply it in volume as original equipment. In fact, only Jensen would actually take the bait, and they fitted it to their exclusive and very expensive FF model, of which just 320 were built between 1967 and 1971.

The Ferguson Formula traction control system was nothing less than a full-time four-wheel-drive system with anti-lock brakes. It depended on a chain-driven transfer box which took the drive from the main gearbox and turned it round to take it forward to the front wheels. Allied to this was the Dunlop Maxaret anti-skid braking system (as it was known at the time), which had been developed from the systems then in use for the undercarriages of aircraft to prevent slewing under braking during landing in wet conditions.

In October 1969, Harry Ferguson Research sold its four-wheel-drive patents to GKN, who chose an interesting method of marketing the system. They decided to convert a number of cars as demonstrators in the hope that their manufacturers would be suitably impressed by the system's merits once they had tried the converted cars. One manufacturer which was impressed was Ford, which commissioned a small number of Mk.IV Zephyrs for Police use, but then took the trial no further. Among the other manufacturers which GKN targeted was British Leyland, and the car they chose for conversion was the Triumph Stag.

GKN accordingly bought a pair of Triumph Stags, one with automatic transmission and one with a manual gearbox, and arranged for Harry

Above: The only outward sign that there is anything unusual about the 4x4 Stags is this discreet bonnet bulge, which had to be added when the engine was raised slightly to clear the additional transmission components.
Below: This is the automatic 4x4 Stag, a 1971 car. It has been updated with later five-spoke alloy wheels and with bright sill covers.

Opposite page
Completed in 1984, Tickford's comprehensive reworking of a 1975 Stag gave it a strongly aggressive look, together with improved handling and a luxurious interior. The contemporary Aston Martin Vantage was a clear influence. (MRP)

Ferguson Research to convert them to Ferguson Formula four-wheel drive with anti-skid brakes during 1972.

At the forward end of the gearbox, a transfer gearbox mounted on the passenger side of the car took the drive forward to the front wheels, giving the 40 percent/60 percent front-to-rear torque split which Ferguson had found to be ideal for optimum control in all conditions. The original suspension layout was maintained, but room was made for an offset differential by raising the engine slightly. The new front drive-shafts were fabricated, probably from cut-down BMC 1100/1300 items.

None of these changes was visible from outside the car, but their effects were. The bonnets had a neat and carefully fabricated bulge — necessary to clear the raised engine — while there was another bulge inside, this time on the side of the transmission tunnel to accommodate the transfer box. Just ahead of the gear lever was a small rocker switch which could switch off the Maxaret anti-skid braking system, as this did not suit certain types of surface.

How seriously Triumph took the GKN/Ferguson exercise is not clear. However, by the time the prototypes were ready, they had become painfully aware of the Stag's shortcomings and

were probably far too concerned with sorting them out to consider a major modification to the design. The automatic 4x4 Stag was soon relegated to the function of runabout for the wife of GKN's Chairman, and was then sold to a Birmingham garage proprietor who kept it in Spain as holiday transport. It now belongs to Steve Barratt, the proprietor of Triumph parts specialists S.N.G. Barratt. The manual-transmission car also still survives, and is in private ownership.

The Crayford estate car

Crayford Auto Development of Westerham was well-known in the late 1960s and early 1970s for its conversion work on standard production cars, turning saloons into convertibles, estate cars or hatchbacks. David McMullan, a Director of the company, confirmed several years ago that a Triumph Stag was converted into an estate car as a development exercise but that the project was taken no further.

It is not clear what happened to this car. In 1982, an advertisement appeared in the *Thames Valley Trader* for a 1975 car which was described as a Crayford Stag estate, and the rather poor photograph which went with the advertisement strongly suggests that it was indeed a two-door estate car based on a Stag. It was described at the time as having Recaro seats and Minilite wheels.

This car is not to be confused with a small number of other conversions, carried out by a company at Weston-super-Mare, where a Stag V8 engine was transplanted into the bodyshell of one of Triumph's 2000/2500-series estate cars.

The Tickford Stag

During 1984, the Tickford coachbuilding division of Aston Martin completed conversion of a 1975 Stag to a customer's individual specification. The customer asked for the car to be made faster, more aggressive, more luxurious and more eye-catching.

In order to improve performance, the cylinder heads, camshafts and manifolds of the Stag's V8 engine were modified, and a less restrictive stainless steel exhaust system was designed and fitted. For longevity, the engine was balanced and equipped with an additional oil cooler. The results were 200bhp and a top speed in excess of 120mph.

Handling was also improved by lowering the suspension and fitting uprated dampers, together with six-inch wide alloy wheels and 225/60 Pirelli P6 tyres. To cover these, the wheelarches were flared slightly and given an additional lip.

The changes to the body styling were largely inspired by the contemporary Aston Martin Vantage. The grille was blanked out with a plain panel, and air ducting for the radiator was incorporated in a deep front spoiler, which was hand-made in aluminium. An aggressive-looking 'power bulge' air intake was also added to the bonnet, while an undertray at the rear was intended to improve the lines. To complete the aggressive look, the car was painted black, and the chromework was finished in semi-matt black to tone in.

Inside the car, a luxurious finish had been created by extensive use of burr walnut, together with Connolly hide and Wilton carpet. A high-power stereo system was fitted, and there were electrically adjustable door mirrors and tinted glass.

Tickford's press release about the car, issued in March 1984, expressed the hope that other Stag owners might be inspired to seek similar conversions. However, the Tickford Stag seems to have remained unique.

Engine transplants

British Leyland never managed to make the Stag engine wholly reliable during its production life, and it was left to aftermarket specialists to develop cures for the V8's maladies. By the mid- 1980s,

Top: The Rover V8 engine has always been the favoured replacement engine for Stags. However, with its standard inlet manifold and SU carburettors, it is too tall to fit beneath the Stag's bonnet...
Above: ... with the result that owners have to fabricate a suitable 'bulge'. This one was made from the bonnet bulge of a Ford Capri.

Above: Some Rover V8 transplants do not need the bulge, however. A lower line has been achieved in this conversion, which uses an Offenhauser manifold and a Holley carburettor, both primarily designed to increase the performance of the Rover engine.

these specialists had been so successful that a properly-treated Stag engine could be as reliable as it always should have been from the start; but in the late 1970s and early 1980s many owners played safe by replacing wrecked engines in their Stags with engines of known reliability.

This trend was perhaps inevitable. By the end of the 1970s, many older Stags had passed on to second and third owners, who probably ran the cars on a more limited budget than their first owners had enjoyed. Servicing was skimped, regular maintenance overlooked, and the result was that the engine problems which had plagued the car during its production life developed into an epidemic. As their owners were in many cases unable to afford the cost of a new Stag engine or reasoned that even a new Stag engine might suffer from the same reliability problems as the old, they turned elsewhere for replacement engines.

Obviously, for owners like these, the alternative to the Stag V8 had to be cheap, readily available, and durable. In Britain, there were two obvious choices: the Rover 3528cc V8 and the Ford 2994cc V6. Both had been built in large quantities since their introduction (1966 for the Ford, 1967 for the Rover) and were readily available for small money from scrapyards. Both also offered the right sort of power output, and both could be tuned for extra performance without too much difficulty.

Exactly who put the first Ford V6 or Rover V8 into a Stag is something which will probably never be established for certain. One way or another, though, these two ideas caught on in a big way and a number of small companies started converting Stags. Some of them carried out well-engineered conversions, but others were little more than back-street garages hoping to make some easy money. The cars converted by such companies, and many converted at home by DIY mechanics, often have serious shortcomings which border on the dangerous.

The all-alloy Rover V8 engine has always been the most popular 'transplant' unit for the Stag. Originally designed by General Motors in the USA for use in the 'compact' Buicks, Oldsmobiles and Pontiacs of the early 1960s, it was taken out of production when new manufacturing technology enabled GM to make cast-iron engines which were

Post-factory modifications

Right: The other popular transplant engine is the Ford V6, usually in three-litre Essex form, although the 2.8-litre and 2.9-litre Cologne types can also be fitted. In this picture, the air cleaner has been removed to show the installation more clearly.

Below: Some Stag conversions utilise Triumph's own six-cylinder engine, around which the earliest prototypes were designed. This is the 2½-litre twin-carburettor version. Performance is not inspired.

physically no larger than the Buick-Oldsmobile-Pontiac V8 but which did have greater swept volumes. Rover bought the manufacturing rights to the alloy V8, redeveloped it to suit British manufacturing practices and motoring conditions, and introduced it in 1967. Most engines in Stags have been the 143bhp version from the Rover P6 3500 or the 155bhp version from the SD1 3500, though some owners might have mistakenly bought low-compression Land Rover or Range Rover versions, which offer less power. Right at the top end of the scale are the 190bhp fuel-injected version from the Rover SD1 Vitesse and the later 3.9-litre and 4.2-litre high-compression Range Rover engines, with 185bhp and 200bhp respectively. In addition, large-capacity and tuned versions have been available from specialists for many years now.

As the Rover V8 is a very understressed engine, it is also very durable. Pre-SD1 versions will not rev as high as the Stag V8, however, which limits top speed even though power is similar to the Stag engine's. The Rover engine offers quite a lot more torque than the Stag V8, and can give better acceleration. However, the Rover V8's height can cause installation problems in the Stag, and its light weight can cause handling and braking problems. All the necessary braking and springing modifications were carried out by the more reputable firms which carried out conversions themselves, but home-converted V8 Stags often lacked these very necessary final touches.

The height problem is confined to those versions of the Rover engine fitted with twin SU carburettors. In order to maintain adequate clearance under the bonnet, many early conversions had to incorporate a bonnet bulge of some sort. Not all of them were fabricated either neatly or skilfully!

Underbonnet clearance does not present a problem with the Ford V6, which fits neatly into the Stag's engine bay. However, both power and torque outputs are lower than those of the Stag V8, with the result that untuned versions of the Ford V6 tend to make the Triumph feel rather underpowered. The V6 is also nowhere near as smooth and refined

Left: *Richard Lane's superbly engineered Stag has been extensively rebuilt for competition in sprint races, but still looks remarkably standard.*

Opposite page
The engine in Richard Lane's Stag is a 400bhp, 4.5-litre British Leyland Motorsport V8, with four twin-choke Dellorto carburettors.

as the Triumph V8, and converted cars thus lose some of the Stag's essential appeal.

As the Stag was originally designed to take Triumph's straight-six engine, it is not surprising to find that the engine bolts straight into the Stag's engine bay with the minimum of difficulty. In fuel-injected form, the 2498cc six-cylinder engine gives up to 150bhp, which is well up to the Stag V8's 145bhp. Torque, however, is not as high, with the result that flexibility is lost; and the straight-six's exhaust note is very much less stirring than the rumble of the standard V8. The straight-six engine also exists in 99bhp carburetted form and in smaller-capacity 1998cc form, but both of these engines give very disappointing performance characteristics in a Stag.

Over in the USA, conversions have been made using small-block Chevrolet V8 and even Volvo 2.7-litre V6 engines. However, these have been done very much on an individual basis, and no company has ever marketed a fully-developed conversion kit for these engines. The durability and success of such conversions therefore depends very much upon the skill of those who carried them out.

Even though the Stag's own V8 engine can now be made as reliable as anyone could wish for, the conversion scene is far from dead. The tuning potential of Ford V6, Rover V8 or small-block Chevrolet V8 engines continues to attract those who want to make the Stag into a high-performance car by modern standards. In Great Britain, the Stag Owners' Club has a special section to cater for enthusiastic owners who run modified cars.

Stag pick-ups

Triumph never built any Stag-based pick-ups, but a small number of owners have carried out their own conversions. As a two-door car with a long rear deck, the Stag lends itself quite readily to such a conversion, and the position of the fuel tank below the boot floor allows the full length of the vehicle behind the front seats to be used as a load deck without major modification. The first Stag pick-up to be built was probably the one which Hart Racing Services used as a parts collection and delivery vehicle; but other Stag specialists have followed suit since.

Even though the Stag does make quite a good pick-up if a cab roof is created out of the front half of a hard top and a rear window is fabricated, such vehicles have amusement value only and their fuel consumption makes them uneconomical as working commercial vehicles.

Racing Stags

No Stag ever raced in British Leyland's colours, as the company believed it had more competitive cars to use in motorsport. Nevertheless, a racing Stag was built in 1979, two years after the car had gone out of production.

The builder was not British Leyland's competitions department, but Tony Hart of Hart Racing Services, then a newly-established company specialising in the care and maintenance of Stags. Letting his enthusiasm run away with him during an interview with a journalist, Tony Hart claimed that his company was considering building a racing Stag to prove that a Triumph V8 engine could be as reliable as any other. When that comment found its way into print, Hart realised he had better do something about it!

As a result, the company gutted an early MkI Stag, lightened it where possible, and fitted it with wide wheels to take racing tyres. In original form, it ran a standard engine and this, as predicted, proved thoroughly reliable during its first season of racing. However, the Stag was not competitive when running against the much more extensively modified cars then allowed under the Modsports rules, and so Hart Racing Services made some further improvements for the car's second season of competition.

The Stag was extensively redeveloped with lightweight fibreglass body panels, modified suspension, and a tuned engine which allegedly developed 220bhp. The power increase had been achieved by means of four Weber carburettors, big valves in ported and polished heads, special Piper camshafts, and four-branch extractor manifolds. For safety, the engine was also strengthened. It then performed faultlessly for two seasons' racing, although the car was not seriously competitive.

After an accident, the car was rebuilt again with an even more powerful engine of 270bhp, but it was still not really a front-runner. The car has not been raced seriously since the mid-1980s demise of the ModSports category in which it competed, but is still owned by Hart Racing Services.

A Stag was also rallied in a number of British club and regional events during the 1980s. The car was owned by Geoff Staples, who had earlier rallied a modified Triumph 2.5 PI, and who had transferred its mechanical elements into the bodyshell of a crashed Stag. The car is in fact still registered as a Triumph 2.5 PI, as the licensing authorities consider that it has simply been re-shelled! The rallying Stag had a 240bhp Rover V8 engine and 5-speed gearbox, and originally used a Triumph 2000 differential. However, repeated differential and half-shaft failures eventually prompted a change to a Jaguar independent rear end. Other non-Stag elements in the car included 13-inch wheels (chosen because of the ready availability of appropriate tyres) and a Triumph TR6 steering rack.

A further racing Stag was built in the early

1990s by Richard Lane, owner of a Brighton-based company specialising in Stags. The car was awarded highest marks for its engineering quality in *Cars and Car Conversions* magazine's Castrol Converted Car of the Year competition in autumn 1992. As figures show, it had been built primarily for sprint events: from rest to 60mph took just 4.5 seconds, 0-100mph took 9.5 seconds, and the car's mean maximum speed was 156mph.

At the heart of Richard Lane's Stag was a 4,541cc British Leyland Motorsport Tuscan Challenge full race engine (a derivative of the Rover V8), which would rev as high as 7,500rpm. This drove through a British Leyland Motorsport five-speed racing gearbox to a Quaife torque-biasing differential installed in a narrowed Rover SD1 rear axle. Suspension and brakes had, of course, been fully re-engineered to cope with the car's astonishing performance. Perhaps of greatest interest to Stag enthusiasts, however, was the fact that the car looked almost standard.

Overleaf
The many improvements wrought in the Stag since production have gained it a very strong following. A 1993 reader's poll conducted by the widely-read monthly Your Classic *earned it first place — for the second consecutive year — as the best all-round classic up to a value of £10,000, against strong competition from the Jaguar MkII, Morris Minor, Mini-Cooper and others.*
(Reproduced by kind permission of Your Classic*)*

YOUR CH
Classics of the

Which, in your opinion, is the best all-round classic car? Which is the best-looking, and which do YOUR CLASSIC readers actually own? We reveal all...

THE 1993 READERS' POLL results are in. There are a few unexpected surprises, a couple of new entries in the 'charts' and some interesting changes of position.

With the car market in its current state, prices of exciting cars are back in the realms of reality, so our cut-off figure of £10,000 includes some exotic machinery. For the classic car enthusiast things have never been better; machines which were previously mere dreams are now firmly within grasp and affordable. So, here's how you voted.

Best all-round classic for £10,000

The odds-on favourite has romped in for the second year running. The Triumph Stag gained 14 per cent of your vote in this category as the most

CHOICE of the year '93

READERS' POLL

MkI Golf GTI establishes its classic status at No 11 in Best all-round category. E-type thrust its way into 15th place

The '70s are back; sporty Scimitar is at No 5

desired all-round classic car. With its V8 engine, sculpted Michelotti styling and good specialist back-up, this big British sports car is ahead in the stakes.

But the biggest change in this year's poll, when compared with last year, is the Jaguar's second place. It came second last time around, too, but this year the Stag only just pipped it to the post. The MkII is gaining in popularity (with 12 per cent of your vote) and, as the prices for Jaguars continue to become more realistic, the lure of the big cat is obvious. With that lusty double-overhead-cam six-cylinder engine and curvacious feline body, the Jaguar's rakish image is very attractive. Smart buyers go with not quite the image of the MkII, but is cheaper to buy. So the Jag could be on its way to beating the Stag. We'll have to wait and see what happens next year.

Third place goes to the ever-popular MGB (fifth last year) and the Morris Minor is at number four. The top 10 places are well and truly the preserve of traditional British classics, but some newcomers are pushing through from the back of the field. Following the very popular Golf GTI MkI *Fact File* in the January 1993 issue, this modern classic is voted 11th as best all-round Classic for under £10,000. A sign of the times and the fact that some modern cars are now becoming sought-after. We had a long, hard think about whether it is time to include cars like the Golf, but your votes confirmed that we did the right thing.

Another high climber is the Porsche 911, which you voted into 13th place in the best all-round category – a sporting machine no driving enthusiast can ignore. The E-type Jaguar just scraped in 15th; another exotic which is becoming a more affordable reality all the time. Get in while the going's good.

Best looking classic to buy for £10,000

Here, the Jaguar MkII beats the Stag by a nose, with the E-type coming in third and the Triumph fourth. A big surprise is that MGA scooped up 5.8 per ▶▶▶

Surprise climbers: MGA and 911

Top of the Pops

	1992	1993
1	Triumph Stag	Triumph Stag
2	Jaguar MkII	Jaguar MkII
3	Morris Minor	MGB
4	Mini-Cooper	Morris Minor
5	MGB	Reliant Scimitar
6	Rover P5	Mini-Cooper
7	Reliant Scimitar	Rover P6
8	Jaguar XJ6	Triumph TR6
9	Triumph Vitesse	Rover P5B
10	VW Beetle	MGB GT V8

Peak Performance Co. Ltd.
TRIUMPH STAGS SALES ✦ SERVICE ✦ PARTS

PEREGRINE WORKS ✦ 210-214 LAMPTON ROAD ✦ HOUNSLOW ✦ MIDDLESEX TW3 4EL
081-570 0477 — 081-577 1350 — 081-570 7037 (FAX)

THE HEATHROW STAG CENTRE   Established 1978

As one of the country's leading Stag specialists we continue to offer a comprehensive service to all Stag owners by way of our fully equipped workshops and highly skilled staff offering routine servicing to full engine/gearbox/final drive rebuilds, brake and suspension overhauls, tyre and exhaust fitting, interior trimming, new soft top fitting, burglar-alarm installation, in fact **any job you need on your Stag.**

- ❖ **Our paintshops** are constantly renovating these cars from minor crash repairs to full restoration, often trebling their value.
- ❖ **Our parts department** holds a comprehensive range of new, used and reconditioned Stag parts, all at competitive prices.
- ❖ **Finally, our sales department** normally carries at least **15 Stags for sale,** all serviced, MOT'd and with a 3-month guarantee prior to collection, from £3,000 upwards. If we haven't got the Stag you really want in stock, we can quite often locate it from our large customer base.

Phone for details and current stock list.
STAGS ALWAYS PURCHASED FOR THE BEST POSSIBLE PRICES OR SOLD ON COMMISSION IN OUR SHOWROOM.
081-577 1350 — 081-570 0477 — Fax 081-570 7037

ALDRIDGE
TRIMMING
Specialists in Car Trimming
BRITISH MOTOR HERITAGE APPROVED

THE COMPANY
Occupying 20,000 square foot factory in Wolverhampton town centre, employing 30 skilled trimmers, machinists, cutters and fitters.

FITTING
Because we personally trim well over 100 cars a year, we know the problems encountered with fitting new trim to your car, and are quite happy to give assistance.

MANUFACTURING
As an original equipment supplier to the motor industry, we manufacture 95% of all parts we supply, from leather seat cover to a sun visor.

QUALITY
Our company has British Motor Heritage approval, and all parts supplied to you carry a money back guarantee.

ENQUIRIES
For further information on whole or part trim kits, please do not hesitate to call or fax.

ST. MARKS ROAD, CHAPEL ASH, WOLVERHAMPTON WV3 0QH

   Telephone: (0902) 710805   Fax: (0902) 27474

Appendix A — TECHNICAL SPECIFICATIONS

TYPE DESIGNATION Triumph Stag
DRIVE CONFIGURATION Front engine, rear-wheel drive.
ENGINE
    Type: 90° V8 with five-bearing, cast-iron block and aluminium alloy cylinder heads with overhead valves.
One overhead camshaft per cylinder bank.
    Capacity: 2997cc (182.9cu in)
    Compression ratio: LF...HE engines below 20001: 8.8:1
LF...HE engines, 20001 and above: 9.25:1
LE...UE engines, below 20001: 8.0:1
LE...UE engines, 20001 and above: 7.75:1
    Bore and stroke: 86mm x 64.5mm (3.39in x 2.54in)
    Maximum power: LF...HE engines, below 20001: 145bhp DIN @ 5500rpm
LF...HE engines, 20001 and above: 146bhp DIN @ 5700rpm
LE...UE engines, below 20001: 127bhp @ 6000rpm
LE...UE engines, 20001 and above: 127bhp @ 5500rpm
    Maximum torque: LF...HE engines, below 20001: 170 lbs/ft @ 3500rpm
LF...HE engines, 20001 and above: 167 lbs/ft @ 3500rpm
LE...UE engines, below 20001: 142 lbs/ft @ 3200rpm
LE...UE engines, 20001 and above: 148 lbs/ft @ 3500rpm
    Fuel system: Engines numbered below 20001:
two Zenith-Stromberg 175CD caburettors
Engines numbered 20001 and above:
two Zenith-Stromberg 175 CDS (E) V carburettors
MANUAL GEARBOX Four forward speeds, all-synchromesh, driven through a 9in diameter single dry plate clutch.
Laycock A-type overdrive operating on third and top gears optional prior to October 1972; standard from October 1972 to January 1973 except in certain export markets.
Laycock J-type overdrive operating on third and top gears standard from February 1973 except in certain export markets.
Ratios: 2.995:1, 2.1:1, 1.386:1, 1.0:1; reverse 3.369:1.
Overdrive 3rd (pre-February 1973) 1.135:1, overdrive top 0.82:1.
Overdrive 3rd (post-February 1973) 1.10:1, overdrive top 0.797:1.
AUTOMATIC GEARBOX
    Pre-October 1976: Borg Warner type 35 with three forward speeds, driven through a torque convertor.
Ratios 2.39:1, 1.45:1, 1.0:1; reverse 2.09:1.
    Post-October 1976: Borg Warner type 65 with three forward speeds, driven through a torque convertor.
Ratios as for type 35.

FINAL DRIVE:	3.7:1.
DIMENSIONS	
Length:	14ft 5.75in (4420mm)
Width:	5ft 3.5in (1612mm)
Height:	4ft 1.5in (1258mm) with hood erected
Wheelbase:	100 in (2540mm)
Track:	Front: 52.5in (1330mm). Rear: 52.9in (1340mm)
SUSPENSION	
Front:	Independent with MacPherson struts, lower links, anti-roll bar and telescopic dampers.
Rear:	Independent with semi-trailing arms, coil springs and telescopic dampers.
STEERING	Power-assisted rack-and-pinion
BRAKES	Servo-assisted split-circuit system with tandem master-cylinder. 10.6in diameter discs at the front; 9in diameter drums at the rear.
WHEELS AND TYRES	
Pre-October 1975:	5.5J pressed-steel disc wheels with four-stud fixing and push-on trims. Centre-lock chromed wire wheels available until mid-1975 only. Five-spoke alloy wheels optional from 1973. 185HR14 tyres (Michelin XAS fitted as standard to January 1973; Michelin XAS or Avon alternatives from February 1973).
Post-October 1975:	5.5J five-spoke alloy wheels standard. 185HR14 tyres (Michelin XAS or Avon fitted as standard).
WEIGHT	2981 lbs (1355kg) with hard top.

E. J. WARD MOTOR ENGINEERS
Triumph Stag Specialists, Mechanical & Bodywork Repairs
Full Restorations. Stags for Sale

66 Jarvis Street, Leicester
Tel: (0533) 519775
Tel: (Eves & Weekends) (0533) 782075

ESTABLISHED IN 1979

Membership of the Stag Owners Club now exceeds 6,000. The Club is run by a National Committee and has over 46 local areas where members can meet to increase their enjoyment of Stag motoring.

Local, National and International Meetings are held throughout the year. Our annual National Day attracts over 700 Stags and is held at different locations around the country.

Spare parts are readily available thanks to a host of specialist parts suppliers, many of whom are remanufacturing unobtainable or hard to find items.

There are special insurance terms with a valuation service for members taking part in the schemes.

An illustrated magazine of over 60 pages is sent to all members each month. It features articles on Stag maintenance, local area news, details of forthcoming events, members letters, technical queries complete with answers from the Club's consultants, advertisements and a variety of other information.

A wide range of car badges, T-shirts, key fobs, hats, tankards, magazine binders, etc, are available direct from our Accessories Secretaries.

For membership details please send a S.A.E. to:
The Membership Secretary, Stag Owners Club,
53 Cyprus Road, Faversham, Kent ME13 8HD

STAG IN SOUTH WALES

AUTOMOTIVE SERVICES
THE WORKSHOPS,
CRAWFORD ST.
NEWPORT, GWENT
0633 253205

SOUTH WALES LONGEST ESTABLISHED TRIUMPH SPECIALIST

FULL RESTORATION & REPAIR FACILITIES
LOW BAKE OVEN, BODY JIG & TILT FACILITIES
SHOT BLASTING & POLISHING
ENGINE & GEARBOX OVERHAUL AND REPAIRS
REPAIRS & RESTORATIONS FOR ALL CLASSIC CARS

FOR TOP QUALITY WORKMANSHIP BY CRAFTSMAN AT REASONABLE PRICES CONTACT US NOW.
FREE ADVICE AND ESTIMATES

Appendix B — VEHICLE IDENTIFICATION

COMMISSION NUMBERS

The Commission Number corresponds to what used to be called the Chassis Number on cars which had separate a chassis and body and to what is today called the VIN (Vehicle Identification Number). It is found on a plate attached to the left-hand-side door shut pillar.

All Commission Numbers are prefixed with the letters LD (or LE for US Federal models). Suffix letters BW indicate automatic transmission, and O indicates manual transmission with overdrive. A typical example would therefore be LD 31205 BW. The sequences for Stags were:

LD 1-3900	1970-1971 models
LD/LE 10001-14158	1972-season models
LD/LE 20001-25396	1973-season models
LD 30001-36685	1974 and 1975-season models
LD 40001-45722	1976 and 1977-season models

Some Stags were also built in the sequence LE 7500-9000. These were probably all US Federal vehicles, but build records have not survived and it is therefore impossible to be certain how many cars there were. The numbers in the sequences listed above provide a total of 25,861 cars; the totals (see Appendix C) of 25,877 cars delivered and 25,939 produced suggest that between 16 and 78 cars might have been built in this additional sequence.

ENGINE NUMBERS

On early engines, the Engine Number is stamped on a cast boss between the two rearmost spark plugs on the left-hand cylinder bank. On later engines, it is stamped at the back of the cylinder block, between the distributor and the bellhousing. Engine Numbers are prefixed with the letters LF (or LE for US Federal variants), and end with the letters HE (or UE for US Federal variants). Federal engines had their own numbering sequence. A typical example would therefore be LF 11246 HE. The sequences were:

LF 1 HE upwards	1970-1971 models
LE 10001 UE upwards	1972-season Federal models
LF 10001 HE upwards	1972-season non-Federal models
LE 20001 UE upwards	1973-season Federal models
LF 20001 HE upwards	1973-season non-Federal and 1974-1977 models

BODY NUMBERS

A small plate attached to the front crossmember of the body beside the left-hand bonnet hinge gives the Body Number. These begin at 001LD, and do not match the commission numbers (although numbers in most instances will be close to one another).

GEARBOX NUMBERS

There are two different types of Gearbox Number. Manual gearboxes have a Triumph serial number stamped on them, beginning at LD1. This sequence was used for all gearboxes except those on US Federal cars: 1972-season US Federal gearboxes are numbered in a series beginning at LE 10001, and 1973-season US Federal gearboxes begin at LE 20001. Automatic gearboxes have a Borg Warner serial number stamped on a purple plate (type 35 gearbox) or a green plate (type 65 gearbox) attached to the gearbox casing.

AXLE NUMBERS

Axle Numbers are stamped on the underside of the rear axle casing. They are all in the same series, which begins with LD1.

Appendix C — PRODUCTION FIGURES

Season	Home Market	Export	Total	
1970	358	30	388	Manual
	342	10	352	Automatic
	700	40	740	*Total*
1971	1,291	1,221	2,512	Manual
	699	690	1,389	Automatic
	1,990	1,911	3,901	*Total*
1972	2,015	432	2,447	Manual
	1,490	567	2,057	Automatic
	3,505	999	4,504	*Total*
1973	1,511	1,138	2,649	Manual
	1,683	1,176	2,859	Automatic
	3,194	2,314	5,508	*Total*
1974	1,327	405	1,732	Manual
	1,279	431	1,710	Automatic
	2,606	836	3,442	*Total*
1975	935	440	1,375	Manual
	1,051	472	1,523	Automatic
	1,986	912	2,898	*Total*
1976	N/A	N/A	N/A	Manual
	N/A	N/A	N/A	Automatic
	2,466	644	3,110	*Total*
1977	501	112	613	Manual
	871	352	1,223	Automatic
	1,372	464	1,836	*Total*
Total 1970-1977	17,819	8,120	**GRAND TOTAL** **25,939**	

Note: *These figures are used by kind permission of Anders Clausager at BMIHT. They are based on build records and do not include those cars built in the LE 7500-9000 sequence (see Appendix B). They differ from totals first given by Graham Robson in the February 1978 issue of Thoroughbred and Classic Cars. These totals, based on delivery records, were 25,877 (overall), broken down into 19,097 (home) and 6,780 (export). The reason for the discrepancy is not clear.*

Key dates and figures in the Stag production run are as follows:

First off-tools car	LD 1	14th November 1969
Second off-tools car	LD 2	12th February 1970
First line-built car	LD 3	13th March 1970
First car built in 1971	LD 748	1st January 1971
Last first-sanction car	LD 3900	9th November 1971
First second-sanction car	LD 10004	4th November 1971
First car built in 1972	LD 10760	3rd January 1972
Last second-sanction car	LD 14158	10th November 1972
First third-sanction car	LD 20001	17th October 1972
First car built in 1973	LD 21087	2nd January 1973
Last third-sanction car	LD 25396	11th October 1973
First fourth-sanction car	LD 30001	14th September 1973
First car built in 1974	LD 31153	4th January 1974
First car built in 1975	LD 34617	2nd January 1975
Last fourth-sanction car	LD 36685	18th September 1975
First fifth-sanction car	LD 40001	4th September 1975
First car built in 1976	LD 40809	5th January 1976
First car built on Innsbruck line	LD 42366	4th August 1976
First car built in 1977	LD 43908	4th January 1977
Last Stag	LD 45722	24th June 1977

Notes:
1. Build sanctions occasionally overlapped; thus, cars with the specification agreed for the third sanction, for example (i.e. Mark II Stags), started coming off the lines before orders for second-sanction cars had been fulfilled.
2. Until August 1976, Stags were built in batches on the TR6 production lines. When TR6 production ended, Stag production was transferred to the Innsbruck (2000/2500 saloon) production lines.
3. Stags were built in small batches, and the rate of production varied. It is therefore impossible to date the build of a particular car accurately with reference to the list above. Build dates of individual Stags may be obtained from the British Motor Industry Heritage Trust, with whose kind permission this list is reproduced. There is a charge for BMIHT's Production Trace Service.

THE NORTHANTS STAG CENTRE

TELEPHONE: 0933 442299

FAX: 0933 442279

UNIT 99/100, THE LEYLAND COMPLEX
IRTHLINGBOROUGH ROAD
WELLINGBOROUGH
NORTHANTS NN8 1RT

THE COMPLETE STAG SPECIALIST

At the Northants Stag Centre we pride ourselves on the quality of both our parts and our service. We may not always be the quickest or the cheapest but we like to think we're the best. We offer an honest and genuine service. We know most of our customers by their first names and most of their cars better than they do! Our staff are friendly, courteous and, above all, helpful. If we don't stock it we can usually locate it within a day or two.

Our workshop is always busy with several restorations on the go at any one time. All our customers are welcome to come and talk to our staff and view the standard of our work for themselves. We would be only too happy to discuss your requirements with you. We offer a complete service, whether it be bodywork, mechanical or trimming. We can recondition your own units (not always the cheapest way) or supply reconditioned units off the shelf. Come along for a chat and a cup of coffee (you might have to make it yourself) and give your car a treat. If you can't visit personally, we offer a fast, efficient mail order service. Next day delivery available from £11.75 including VAT, cheaper rates for non-urgent deliveries.

RESTORING THIS YEAR
BOTH SURVIVING STAG PROTOTYPES
(DAVE JELL'S FEATURED CAR AND PVC 437 G);
AND LD 7, THE SEVENTH PRODUCTION STAG AND PRESS CAR (RVC 428 H)

OPENING HOURS:
MONDAY-FRIDAY 9am-6pm   SATURDAY 9am-5pm
WORKSHOP 9am-6pm   WORKSHOP 9am-1pm
ALL PRICES INCLUDE VAT AT 17½%

PANELS	MAJOR MECH UNIT	TRIM	BRIGHTWORK
Front wings B.L.............£129.25	Short engine (exch).......£641.55	Soft top mohair from £223.25	Front bumper blade recon
Inner arch repair £32.31	Full engine (exch).........£1598.00		(exch) 2 year guarantee £235.00
Inner arch full £76.38	Man gearbox (exch)......£188.00	Front seat covers from £47.00	
Outer sills B.L. £52.88	Overdrive unit (exch).....£170.38		Front plinths (exch) £52.88
Inner sills..........................£52.88	Auto gearbox (exch).....£180.60	T-bar cover......................£29.38	
Front floor pan £29.38	Torque convertor (exch) ..£83.13		Door handle bowls (ea)...£22.33
Rear floor pan..................£41.13	Diff new 6wp (exch)......£352.50	Carpet set (car) from......£105.75	Rear badge holder (ea)....£11.75
Rear seat pan..................£70.50	Steering rack (exch) £97.88		
Rear wing repair £51.38	Rear halfshaft (exch)......£95.18	Boot carpet set	Overrider (new) (ea) £33.31
Rear wing full.................£381.88	Brake caliper (exch) £52.88	(incl mill-board) £64.63	Front or rear straight.......£17.63
Rear valance B.L. £52.88	Front disc (ea) £35.25		F & R horseshoes £17.63
Front valance B.L.............£44.06	Rear brake drum (ea)......£47.00	Door seals (ea) £14.10	
Front apron B.L................£45.80	Cylinder head (exch)		Alloy wheels recon
Boot lid Mk I or II............£203.85	from £240.88		(exch) (ea) £36.43
Full floor pan (per side)..£114.57	*All exchange units carry*		
All other panels available	*our 12-month unlimited mileage warranty*	*All seals and other trim ex stock, fitting service available*	*All brightwork ex stock, re-chroming service available*

LARGE STOCK OF SECOND HAND PARTS AVAILABLE

Appendix D — COLOUR AND TRIM CHANGES

Note: Triumph paint and trim changes are notoriously difficult to determine and this list may therefore not be exhaustive.

1971 model year

Damson	with Black or Saddle Tan trim
Laurel	with Black or Inca Red trim
Royal Blue	with Black or Shadow Blue trim
Saffron	with Black or Saddle Tan trim
Signal Red	with Black or Saddle Tan trim
White	with Black, Inca Red or Saddle Tan trim

All soft tops were black and all hard tops in body colour.
Sills and tail panels were in body colour.
A few very early cars were finished in Wedgewood Blue.
Valencia Blue was used on at least one prototype, and might also have been used on some very early production cars.

1972 model year

Damson	with Black or Saddle Tan trim
Emerald	with Black trim
Pimento	with Black trim
Saffron	with Black trim
Sapphire Blue	with Black or Shadow Blue trim
Sienna	with Saddle Tan trim
White	with Black, Saddle Tan, or Shadow Blue trim

Emerald did not become available until the beginning of the 1972 calendar year.
All soft tops were black and all hard tops in body colour.
Sills and tail panels were in body colour.

1972 model year
(US Federal cars)

Damson	with Black or Tan trim
Imperial Blue	with Black or Blue trim
Jasmine Yellow	with Black or Tan trim
Laurel Green	with Black, Red or Tan trim
Saffron	with Black or Tan trim
Sienna	with Black or Tan trim
Signal Red	with Black or Tan trim
White	with Black, Red or Tan trim

All soft tops were black and all hard tops in body colour.
Sills and tail panels were in body colour.

1973 model year
(Mark I, to Jan 73)

Carmine	with Black or Saddle Tan trim
Emerald	with Black trim
French Blue	with Black trim
Mallard	with Black or Saddle Tan trim
Pimento	with Black trim
Sapphire Blue	with Black or Shadow Blue trim
Sienna	with Black or Saddle Tan trim
White	with Black or Shadow Blue trim

All soft tops were black and all hard tops in body colour.
Sills and tail panels were in body colour.

(Mark II, from Feb 73)

Carmine	with Black or Saddle Tan trim
Emerald	with Black trim
French Blue	with Black trim
Magenta	with Black trim
Mallard	with Black or Saddle Tan trim
Mimosa	with Black or Chestnut trim
Pimento	with Black or Chestnut trim
Sapphire Blue	with Black or Shadow Blue trim
Sienna	with Black or Saddle Tan trim
White	with Black, Chestnut or Shadow Blue trim

All soft tops were black and all hard tops in body colour.
Sills and tail panel were painted matt black.
All cars had side-stripes in Aluminium, Black, or Gold.

1974 model year

Carmine	with Black or Saddle Tan trim
Emerald	with Black trim
French Blue	with Black trim
Magenta	with Black trim
Mallard	with Black or Saddle Tan trim
Maple	with Black or Saddle Tan trim
Mimosa	with Black or Chestnut trim
Pimento	with Black or Chestnut trim
Sapphire Blue	with Black or Shadow Blue trim
White	with Black, Chestnut, or Shadow Blue trim

All soft tops were black and all hard tops in body colour.
Sills and tail panel were painted matt black.
All cars had side-stripes in Aluminium, Black, or Gold.

1975 model year

Carmine	with Beige or Chestnut trim
British Racing Green	with Beige or Black trim
French Blue	with Black trim
Delft	with Beige or Black trim
Java	with Beige or Black trim
Maple	with Beige or Chestnut trim
Mimosa	with Beige or Black trim
Pimento	with Black trim
Topaz	with Beige or Black trim
White	with Beige or Black trim

All soft tops were black and all hard tops in body colour.
Sills and tail panel were painted matt black.

All cars had side-stripes in Aluminium, Black, or Gold.

1976 model year	Carmine	with Beige or Chestnut trim
	French Blue	with Black trim
	Delft	with Beige or Black trim
	Java	with Beige or Black trim
	Maple	with Beige or Chestnut trim
	Mimosa	with Beige or Black trim
	Pimento	with Black trim
	Topaz	with Beige or Black trim
	White	with Beige or Black trim

All soft tops were black and all hard tops in body colour. Sills had bright cover plates and the tail panel was in body colour. All cars had side-stripes in Aluminium, Black, or Gold.

1977 model year	Carmine	with Beige or Chestnut trim
	French Blue	with Black trim
	Delft	with Beige or Black trim
	Inca Yellow	with Black trim
	Java	with Beige or Black trim
	Leyland White	with Beige or Black trim
	Maple	with Beige or Chestnut trim
	Mimosa	with Beige or Black trim
	Pageant Blue	with Beige or Black trim
	Pimento	with Black trim
	Russet Brown	with Beige trim
	Tahiti Blue	with Beige or Black trim
	Topaz	with Beige or Black trim
	White	with Beige or Black trim

All soft tops were black and all hard tops in body colour. Sills had bright cover plates and the tail panel was in body colour. All cars had side-stripes in Aluminium, Black, or Gold.

It is probable that Inca Yellow, Leyland White, Pageant Blue, Russet Brown and Tahiti Blue were available only on the very last cars. They may have replaced certain other colours in the range.

Notes:
1. The Stag Parts Catalogue refers to Ice Blue and Slate Grey as paint options. Both were available on other Triumph models prior to 1973, but it is not certain that they really were available for the Stag.
2. Sales catalogues often refer to Blue upholstery rather than Shadow Blue, to Red rather than Inca Red, and to Tan rather than Saddle Tan.
3. The Stag Parts Catalogue refers to Grey as an upholstery colour on pre-1973 model cars, but there is no evidence that it was ever genuinely available for the Stag.

THE STAG CENTRE

381/2/3/ GEFFRYE STREET, LONDON E2
THE "ORIGINAL" TRIUMPH STAG SPECIALISTS

Our Company has been caring for, and specialising in, Triumph Stags since 1975
We have
FULL WORKSHOP FACILITIES
AVAILABLE FOR MINOR REPAIRS, MAJOR OVERHAULS,
SERVICING OR RESTORATION WORK.
MASSIVE SELECTION OF NEW PARTS:
FULL AND SHORT ENGINES, DIFFS, STEERING,
SUSPENSION, CHROME, BODY PANELS, ETC.
● STAGS ALWAYS WANTED ●
— ANY CONDITION —
SECONDHAND PARTS AVAILABLE

SEND FOR PRICE LIST FOR FAST MAIL ORDER OR PHONE
071-739 7052
FAX: 071-739 1344
THE LONGEST ESTABLISHED "STAG" SPECIALIST
AT THE SAME LOCATION

T.S.C.S.
STAG SPECIALIST

EST. 1983

Telephone (0795) 478951 days — (0795) 424616 eves

We feel it's about time you, as the owner of an ever-more desirable classic car, should be given a wider range of services at realistic prices.

Because restorations are one of our established specialities, we are now able to offer a wide selection of Stags For Sale. Some stripped ready for restoration to customers' specifications, i.e. any colour, manual/automatic, leather/vinyl etc.

WORKSHOP: From a routine service to a major mechanical overhaul
BODYSHOP: From a small dent to a full restoration
TRIM SHOP: From a set of carpets to a retrim (leather, vinyl), soft tops etc.

"TRIUMPH STAGS ALWAYS BOUGHT AND SOLD"
Established 1983 — Kent's Leading Stag Specialist

The original paint and trim colours were recorded on the cars' Commission Number plates by means of a two- or three-digit number code. Some colours were re-coded with a three-digit letter code during the 1977 season. An "H" prefix with the trim number (in theory) indicated leather upholstery. These codes can be interpreted as follows:

		Paint	*Trim*
11	Black		black
17	Damson	maroon	
19	White	white	
23	Sienna	brown	
26	Wedgwood Blue	pale blue	
27	Shadow Blue		pale blue
32	Signal Red	red	
34	Jasmine Yellow	pale yellow/cream	
43	Saddle Tan		mid-brown
54	Saffron Yellow	dark yellow	
55	Laurel Green	dark green	
56	Royal Blue*	very dark blue	
62	Inca Red		red
63	Chestnut		dark brown
64	Mimosa	mid-yellow	
65	Emerald	bright mid-green	
66	Valencia Blue	mid-blue	
68	Slate Grey	dark grey	
72	Pimento Red	bright red	
73	Maple	beige	
74	Beige		beige
75	British Racing Green	dark green	
82	Carmine	dark red	
84	Topaz	orange-yellow	
85	Java	bright green	
92	Magenta	bright purple	
93	Russet Brown	mid-brown	
94	Inca Yellow	bright yellow	
96	Sapphire Blue	deep blue	
106	Mallard	dark blue-green	
116	Ice Blue	bright pale blue	
126	French Blue	light blue	
136	Delft	bright mid-blue	
AAE	Russet Brown	mid-brown	
CAA	Carmine	dark red	
HAB	Java	bright green	
JAC	Ice Blue	bright pale blue	
JAE	Tahiti Blue	bright mid-blue	
JAG	Pageant Blue	bright blue	
NAF	Leyland White	white	

* Royal Blue and the Imperial Blue used on Federal cars were probably the same colour.

FOR THE BEST IN QUALITY PARTS FOR YOUR STAG, CONTACT SOC

SOC SPARES LTD.
5 WHEELER STREET
HEADCORN, ASHFORD,
KENT TN27 9SH

Telephone: 0622 891777
Fax: 0622 891678

TELEPHONE FOR FREE PRICE LIST

Appendix E — PERFORMANCE FIGURES

TRIUMPH STAG

	UK market, manual/overdrive (1971 model)	UK market, automatic (1971 model)
Mean maximum speed (mph)	116	112
Maximum speeds in gears (mph)		
Overdrive top	116	—
Top	113	—
Overdrive 3rd	105	—
3rd	92	113
2nd	61	89
1st	42	54
Acceleration (sec)		
0-30mph	3.5	4.1
0-40mph	5.5	5.8
0-50mph	7.1	7.9
0-60mph	9.3	10.4
0-70mph	12.7	14.2
0-80mph	16.5	18.6
0-90mph	21.8	24.9
0-100mph	29.2	34.5
Standing 1/4- mile	17.1 sec/ 82mph	17.9 sec/ 78mph
Acceleration on the move (sec)*		
10-30mph	3.0 (1)	3.0 (1)
20-40mph	2.8 (1)	3.0 (1)
30-50mph	3.7 (2)	3.7 (1)
40-60mph	4.3 (2)	5.3 (2)
50-70mph	6.0 (3)	5.9 (2)
60-80mph	7.4 (3)	7.4 (2)
70-90mph	9.8 (3)	9.9 (3)
80-100mph	13.3 (o/d 3)	14.8 (3)
Overall fuel consumption (mpg)	20.7	17.2
Typical fuel consumption (mpg)	22**	20
Kerb weight (lb)	2805	2835
Original test published	*Autocar*, 10th June 1971	*Autocar*, 10th June 1971

* Figures given are the best times achieved; the figure in brackets is the gear used.

** This figure is taken from the original *Autocar* test of a pre-production manual/overdrive car published in the issue dated 30th July 1970.

BRIGHTON STAG SPECIALIST

Qualified Mech. Eng. & British Standards Coded Welder
16 Years of Stag Preparation
Standard Cars — Street Modified & Racing Cars
The Rover V8 Specialist

Judged As Builder of Best-Engineered Modified Car Of The Year 1992
by *Cars & Car Conversions* Magazine

Telephone Brighton (0273 747123) For Anything To Do With Stags

C. ROSSER MOTORS

South Wales' leading Stag Renovation and Restoration Specialists
(full body and paintshop facilities)

♦

Benson & Hedges Concours Finalist

Class Winner 1991

Class Winner 1992

♦

Units 9-11, St Katherines Court,
Winch Wen Industrial Estate,
Swansea SA1 7ER

**Telephone: Day 0792 798247,
Evening 0792 776450**

SURREY STAGS

**The Premier Stag Dealership
in the South-East**

We have moved to bigger and better premises and continue to sell top-quality cars with full restoration and workshop facilities.

**Tel: 0428 656377
Fax: 0428 653343**

The Workshop,
9a Midhurst Rd,
Fernhurst, Haslemere,
Surrey GU27 8EE

	UK market, automatic (1975 model)	US market, automatic (1972 model)
Mean maximum speed (mph)	115	112
Maximum speeds in gears (mph)		
3rd	117	112
2nd	89	82
1st	54	54
Acceleration (sec)		
0-30mph	4.2	4.4
0-40mph	6.0	6.1
0-50mph	8.1	8.5
0-60mph	10.7	11.5
0-70mph	14.7	15.9
0-80mph	19.2	21.9
0-90mph	25.7	30.9
0-100mph	35.7	—
Standing 1/4 -mile	18.3 sec/ 77mph	18.5 sec/ 75mph
Acceleration on the move (sec)*		
10-30mph	3.1 (1)	—
20-40mph	3.2 (1)	—
30-50mph	3.8 (1)	—
40-60mph	5.5 (2)	—
50-70mph	6.1 (2)	—
60-80mph	7.6 (2)	—
70-90mph	10.2 (3)	—
80-100mph	15.3 (o/d 3)	—
Overall fuel consumption (mpg)	20.8	—
Typical fuel consumption (mpg)	—	17.9**
Kerb weight (lb)	—	795
Original test published	*Autocar*, 12th February 1977	*Road and Track*, July 1971

* Figures given are the best times achieved; the figure in brackets is the gear used.
** U.S. gallons

THE STAG AND ITS RIVALS

	Max speed (mph)	0-60mph (sec)	30-50mph (sec)	Overall mpg
Triumph Stag	116	9.3	8.1	20.7
Alfa Romeo 2000 GTV	115	8.9	10.9	20.8
Datsun 240Z	125	8.3	9.0	25.7
Ford Capri 3000GT	113	10.3	—	19.3
Lotus Elan +2S 130	121	7.7	8.5	21.0
Mercedes-Benz 280SL	121	9.3	8.8	19.0
MGB GT V8	125	7.7	6.2	19.8
Peugeot 504 V6 coupé	117	9.3	—	18.0
Reliant Scimitar GTE	120	8.7	7.6	21.7

Note: All figures are for cars with manual transmission, except for the Mercedes-Benz 280SL which had a four-speed automatic gearbox. The Ford was by far the cheapest of these cars and the Mercedes-Benz was expensive enough to fall into a completely different market sector.

Give your Stag new Hart!

HART RACING SERVICES know their way to a Stag's heart — everything from a quick pick-me-up to a full overhaul, in fact. The equipment we employ and the experience we've gained make us *the* people to see to give your Stag more tractable performance and driving pleasure.

We can put more power under your bonnet, rejuvenate your power steering and give you easier starting.
Our normal stock includes full or short reconditioned engines, cylinder heads, gearboxes, differentials, driveshafts, distributors and power steering racks — in short, everything a Stag should want.

What's more, our special "HRS" parts include blank main bearing caps to smooth up your big ends, and special gaskets to resurrect overskimmed cylinder heads.
We can also take your used Stag off your hands, or sell you some at really attractive prices from our huge parts stock.

SO IF YOUR STAG NEEDS NEW HEART, COME TO HART!

HRS Garages (London) Ltd
Tel: 081 963 1080   Fax 081 963 0946
(Outside of normal working hours Tel: 0836 202347)
48 GORST ROAD, PARK ROYAL LONDON NW10 6LD

FAVERSHAM CLASSICS

Unit 15b, Upper Brents Shipyard, Faversham, Kent ME13 7DZ

Tel: 0795-539163
Fax: 0795-470518

Meeting all your requirements for pleasurable motoring including:

Servicing

M.O.T. Testing

New Hoods

Crash Repairs

Complete Trimming Service

Trouble-free Stag Engines

Major Unit Renovations

Stags For Sale

Complete Car Rebuilding

Suspension and Brake Rebuilds and Repairs

Manual Gearbox Conversions

Alarms and Security A Speciality

Excalibur Approved Installer

For more information, contact

THE STAG SPECIALIST FOR THE SOUTH-EAST

Appendix F — THE STAG OWNERS CLUB

There can be few single-model clubs able to boast a membership amounting to 25 percent of the car's total production run. Such is the case, however, with the Stag Owners Club which, founded in 1980, currently has some 6,000 members in Britain and several hundred more overseas.

In addition to local, national and international meetings, the club offers discounted insurance terms, a wide range of regalia, and frequent price advantages on parts and services. It also issues a highly professional 60-page monthly magazine which features local regional reports, a valuable technical questions and answers service, details of forthcoming events, news from overseas clubs, and classified and display advertising. Through its close ties with specialist Stag companies, the Club has also been instrumental in arranging the remanufacture of hard-to-find Stag items. And it is especially proud of the part it has played in improving the Stag's security measures.

Current addresses:

The Membership Secretary,
Stag Owners Club,
53 Cyprus Road,
Faversham,
Kent
ME13 8HD.

Stag Club Of America,
P.O. Box 26453,
Tucson,
AZ 85276,
USA.

Stag Club Austria,
R.P. Cavanara,
Leiding 4,
A-2823 Pitten,
Austria.

Stag Club France,
C. Lescure,
32 rue E Jacquette,
60200 Compiegne,
France.

Stag Club Netherlands,
C. Severien,
Groenhof 372,
1186 GM Amstelveen,
Netherlands.

Stag Club New Zealand,
J.M. Cruickshank,
P.O. Box 1233,
Falkland,
New Zealand.

Stag Club Switzerland,
M. Schonenberger,
Zulgstrasse 52,
CH-3612 Steffisburg,
Switzerland.

Appendix G — BIBLIOGRAPHY

Triumph Cars: The Complete Story,
by Graham Robson and Richard Langworth,
published by Motor Racing Publications.
A full history of the Triumph marque, 'from Tri-Car to Acclaim' as the sub-title has it.

Triumph Stag,
by Andrew Morland,
published by Osprey Publishing.
A book of colour pictures of the Stag.

Triumph Stag, 1970-1977: Choice, Purchase and Performance,
by James Taylor,
published by Windrow and Greene Automotive.
A guide to the potential pitfalls of buying a Triumph Stag today.

Triumph Stag, 1970 -1980,
published by Brooklands Books.
A collection of contemporary road tests and other reports on the Stag in Brooklands' well-known archive series.

Triumph Stag Collection, No.1,
published by Brooklands Books.
A smaller volume of reprinted road tests and other material on the Stag.
(Note: Brooklands Books also publish reprints of the factory-issue Stag workshop manual, parts catalogue and handbook.)

Triumph Stag Restoration,
published by Kelsey Publishing.
A collection of articles from *Practical Classics and Car Restorer* magazine detailing the step-by-step restoration of a 1972 Mark I Stag.

Triumph Stag Super Profile,
by James Taylor,
published by Haynes.
A brief overview of the Stag.

Above: Giovanni Michelotti. Triumph discovered the Italian by a happy chance and soon had him on a retainer as consultant stylist. The Stag was born out of a Michelotti styling exercise on a Triumph 2000. (Giles Chapman Collection)

Below: When the Michelotti prototype was first under discussion at Canley, Triumph's Les Moore sketched up some possible adaptations of the design. Note the fastback and sharply cutaway tail of this one.

Above : Prototype X 783 (see commission plate left) was located by co-author Dave Jell, who now owns it and intends to remove the modifications it has acquired over the years. (Garry Stuart photos, by permission of Popular Classics)

Opposite page
Advertising for the new Stag stressed its competitiveness with foreign grand tourers, promising 'true sports car handling with real saloon car comfort… you can overtake the Continentals in style.' The vehicle pictured was the first production car, which was taken to Italy, Switzerland and the south of France to be photographed in surroundings appropriate to its expected market.

Triumph Stag—challenging the Continentals

Alfa and Merc look out! Triumph leap in with the Stag.

It's true the Continentals make some fine cars. Some people think they've got the reputation for really stylish, pacy grand tourers all to themselves.

But with the entry of the Triumph Stag that reputation's due to fade.

The Stag has a big-muscled, 145 bhp, V8 engine. And the zip you get in the lower gears (0-30 in 3.5 seconds) in no way impairs the flexibility you get in top (50-70 in 8 seconds).

You get true sports car handling with real saloon car comfort. The Stag rides so easily and silently it seems to shrink long, tiring roads. Top speed is 118 mph.

The Stag's style is sporty but suave. Whether you have it soft top or hard top, there's no flash and fiddle-di-dee—just clean, beautifully low-lying lines that underscore vivid performance.

But it gives way to a show of luxury inside. It is richly appointed with electrically operated side windows, walnut veneer fascia and console, plush carpeting, cigar lighter, clock, and so forth.

The shaped-to-you, fully reclining front seats, covered in basketweave upholstery, maintain you in comfort however long the journey. And you can adjust them for height, rake, and fore/aft movement while you're seated. The rear bench snugly accommodates a couple of normal size passengers.

And in the Stag you don't so much adjust heating and ventilation as adjust the climate, so variable are the combinations.

The comprehensive range of instruments is compactly set before the driver in a non-glare fascia. And the controls are perfectly grouped for instinctive operation.

Steering is precise, reliable rack and pinion. But power-assisted—to help you take all the fight out of tight turns.

Suspension is independent all round, with anti-roll bar up front. Track is a wide 52½". Wheel rims are 5J with 185 HR radials. So, whatever the road, the Stag's behaviour is as stable and sure-footed as its namesake's.

Direct-acting servo, divided braking system puts the massive authority of 10¼" caliper front discs and self-adjusting rear drums under your foot.

This summer, with the hood down, rollover bar exposed, and twin exhausts ebulliently burbling, you can overtake the Continentals in style.

And for a good two grand less than the price of the Stag's main competitor. Soft top £1,995 17s. 6d. Hard top £2,041 11s. 5d. Hard top with soft top £2,093 15s. 10d. Ex-work and inc. p.t. Overdrive £65 5s. 7d. Automatic £104 8s. 11d.

Standard-Triumph Sales Ltd, Coventry. Tel: O2O3-75511. **TRIUMPH**
Triumph put in what the others leave out.

Opposite page
As can be seen in these two pictures, the first production Stag is still very much alive. It is owned by Steve Barratt, of parts specialists S.N.G. Barratt.

Top right and above: *This well-maintained 1971-model Federal-specification Stag has been given a new hood in light beige material. The original would have been black. Note the bright sill covers, wire wheels and all-amber light units at the edges of the grille.*

Right: *The automatic transmission lever, with its stepped gate arrangement.*

Above and left: This 1974 Stag has been painstakingly restored and cared for by its owner, including under the bonnet.
Below: Stags at the Stag Owners Club's National Rally in 1992. Amazingly, given a production total of just under 26,000, the Club boasts more than 6,000 members.

Opposite page
Middle: A mildly revamped Stag appeared in the autumn of 1975 and the car remained in production until June 1977. Domestic sales revived slightly, but exports were dismally low and Leyland were fully preoccupied with other projects. (Northants Stag Centre)

Right: Not quite standard, perhaps, but still an impressive-looking motor car: this is a 1972 Mark I Stag which has been fitted with the five-spoke alloy wheels of a Mark II model.

Right: Possibly the fastest Stag of them all, Richard Lane's tastefully re-engineered sprint machine has a 4,541cc BL Motorsport Tuscan Challenge full race engine under its bonnet.

Above, left and opposite page, top: Although it has now been Anglicised with UK-specification lights and number plate holder, this Carmine Red 1973 car owned by parts specialists Rimmer Bros is a genuine low-mileage Federal model. Attractive front end styling is shown to full advantage.

Right: Dashboard of a 1973 Federal Stag. The car's lukewarm reception in America and its poor performance in the marketplace came as a huge blow to Triumph.

Below, left: Commission number plate on a Federal specification car; compare with…

Below, right: …the less detailed plate carried by cars for other markets

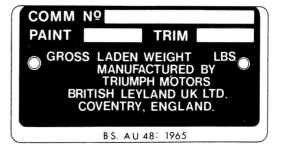

QUALITY IS JUST ONE OF OUR GOOD QUALITIES

Every Triumph is made with a concern for quality that's part of a fifty year pedigree. Quality that expresses itself in obvious ways, and can be seen and appreciated. And quality in a wider sense, not always visible, but every bit as crucial. It's combining both that makes a Triumph a triumph.

But don't take our word for it. Here are just a few of the nice things independent experts have said about just one of our good qualities, engineering, over the past couple of years.

"This outstanding engine..........Outstanding low speed torque is immediately apparent. Not only is the pull strong, almost from idling speeds upwards, but it is also very smooth..........utterly effortless cruising at 90 m.p.h." MOTOR JAN 8/72

DOLOMITE SLANT 4 ENGINE

TRIUMPH DOLOMITE

"There's no other British car quite like it, the combination of performance, luxury, refinement and driver appeal calling to mind some of the (more expensive) wares of BMW, Alfa Romeo, Fiat...?"
MOTOR JAN 8/72

TRIUMPH STAG

"Outstanding impressi of the car, however, was the sheer pleasure of driving behind a multi-cylinder engine. The smoothness and surge of power from the V8 is a truly delightful feeling."
MOTOR MAY

Triumph
Cars that live up to their name

Rover Triumph, British Leyland U.K. Limited, Coventry. Tel: 0203/75511.

Above: Bernard Watson, co-Director of specialists Heathrow Stag Centre, is the owner of this immaculate late-model Stag. The complete restoration took more than 18 months — though it should be pointed out that it was a labour of love, rather than a commercial project, and was carried out in the owner's own time. Originally belonging to the US Embassy in London, the car was stolen but subsequently recovered and was sold to Bernard Watson by its insurers, Norwich Union.

Below: JHC 888 is an example of the 'personalised' Stag, as evidenced by the special wheels, the V8 badge on the front wing (borrowed from a Rover) and — interestingly — the high-back (US spec) MkI seats. This is an early car to which later features have been added. Note the dual coachline and bright sill covers. (SOC)

Opposite page
Significantly, perhaps, this 1974 advertisement finds the Stag sharing the limelight with its contemporary, the Dolomite. Slant 4 and V8 engines suffered from similar manufacturing defects, which rather gives the lie to the loud proclamation of 'quality'.

Opposite page
Many Stag owners today enjoy personalising their vehicles. XCU 58 L is an early Stag which has been resprayed in an attractive non-standard silver and fitted with wire wheels, although not Triumph's own variety. (SOC)

Right: A well-maintained standard-specification Stag — probably one of the last. (SOC)

Right: A carefully restored MkI Stag in Signal Red with black interior. (SOC)

Below: This Federal Stag has extended tailpipe to prevent adverse effect of exhaust fumes on chrome. Note also US rear side marker light, with Stag badge. (D. Bergquist)

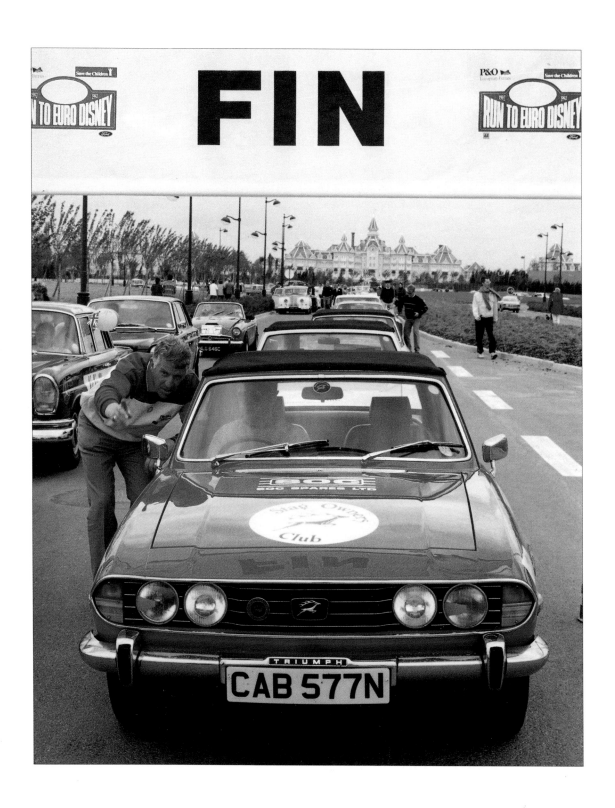

Above: *Stags play an active role alongside other models in classic car events. Here one is pictured at the finish of the 1992 Run To Euro Disney. (Chris Harvey)*

CARDINAL TRIUMPH

TR7 • TR8 • SPITFIRE • 2000/2500 • STAG

PARTS

We can supply from stock all parts that are currently available, as well as some that are not. Our parts department will do their utmost to ensure that you find the parts you need at a competitive price whether new, reconditioned or used spares are required. Our reconditioning is done in-house so we can guarantee first class units at very competitive prices at quality you won't beat! Please phone, fax or write for a quote or price list.

WORKSHOP

As we are true Triumph Specialists we have full workshop facilities for the repair, restoration or general maintenance of your Triumph. As we are constantly fitting the parts we stock we can avoid stocking parts that don't measure up to our strict standards. Yet another way we are able to check the quality of the parts we supply to you. Please telephone to discuss your workshop requirements.

RETAIL AND TRADE

We supply parts to both retail and trade customers and enjoy building a personal customer service that is second to none. Our friendly, expert staff will always take the time to ensure your requirements are met as we want you to use our services time after time. So please don't hesitate to call us if you need advice or parts or just a quote, we are here to help. Export enquiries welcome.

WE ARE STOCKISTS OF FALCON QUALITY STAINLESS STEEL EXHAUST SYSTEMS WITH LIFETIME GUARANTEE

MAIL ORDER

As we accept credit or debit cards the majority of our customers place their orders by telephone or fax. We try to ensure that all our orders are despatched the same day we receive them so if you requested a next day delivery whether you live in Cornwall or Cumbria, a next day delivery is what you get. For none urgent or small orders, 3 day and post delivery is available.

OPEN: Mon-Fri 8.30am - 5.30pm. Sat 9.30am - 2.00pm
Tel: 091 478 5444   Fax: 091 478 4739

CARDINAL TRIUMPH SUPPLIES, CARDINAL HOUSE, HIGH LEVEL ROAD, GATESHEAD, TYNE & WEAR NE8 2AG

PETE EDWARDS STAG WORKSHOP

5 The Courtyard ♦ Thrush Road
Poole ♦ Dorset
Tel: (0202) 747338
Tel: (0202) 731570 (after hours)

THE SOUTH COAST'S OLDEST AND MOST EXPERIENCED STAG SPECIALIST. ALL WORK IS CARRIED OUT ON THE PREMISES.

REPAIRS, SERVICING, BODY RESTORATION AND TRIMWORK. A FULL PHOTOGRAPHIC RECORD IS KEPT OF ALL WORK CARRIED OUT.

☆ STAGED PAYMENTS arranged to suit your budget

☆ WE SPECIALISE IN REBUILDING engines and gearboxes (manual and automatic)

☆ YOUR CYLINDER HEADS can be converted to lead free fuel, or we have exchange cylinder heads in stock

☆ A COMPREHENSIVE RANGE of Stag spares — new, reconditioned or second-hand — is always available

☆ WE ALWAYS HAVE a stock of Stags for sale — please phone for stock list

☆ WE ARE ALWAYS PLEASED TO SEE YOU — pop in any time for a chat

ALL CARS ARE FULLY INSURED WHILST IN OUR CARE

BOSCH

Bosch Technical Instruction

Comprehensive information made easy.

BOSCH

Bosch Technical Instruction

The latest know-how straight from the source

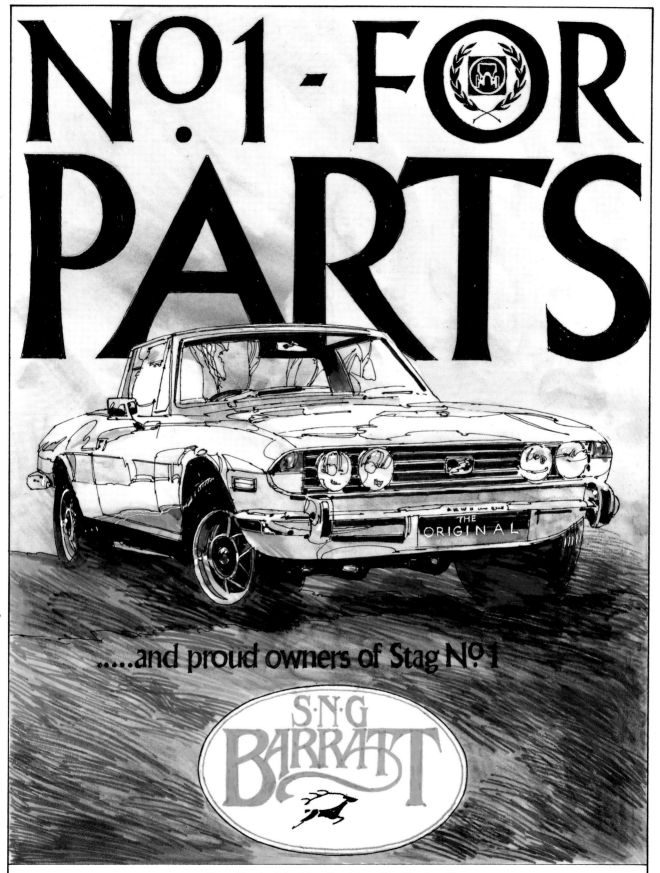